Graphing Calculator Instruction Guide

To Accompany

Calculus Concepts: An Informal Approach to the Mathematics of Change

LaTorre/Kenelly/Fetta Reed/Harris/Carpenter
Third Edition

Iris Fetta Reed
Clemson University

D1452256

Houghton Mifflin Company   Boston   New York

Sponsoring Editor: Lauren Schultz
Developmental Editor: David George
Editorial Associate: Kasey McGarrigle
Production Coordinator: Charline Lake
Senior Manufacturing Coordinator: Florence Cadran
Senior Marketing Manager: Danielle Potvin

Printed in the U.S.A.

ISBN: 0-618-40132-6

23456789-CRS-08 07 06 05

Preface

This *Guide* uses actual examples and applications in the applied calculus text *Calculus Concepts: An Informal Approach to the Mathematics of Change* by Donald R. LaTorre, John W. Kenelly, Iris Fetta Reed, Cynthia R. Harris, and Laurel L. Carpenter. Wherever the technology icon 📖 appears in the text, a new technology technique or an effective way to use your calculator for that particular text material is illustrated in this *Guide*. The technology icon is your clue to refer to the particular section of this *Guide* for your specific calculator that discusses the text section. When there is no reference to a certain section in the text, either there is no new procedure to learn or the necessary techniques have been covered in an earlier section of this *Guide*. Refer to the table of contents to find specific types of applications and the *Troubleshooting* sections in this *Guide* when you have problems.

The use of technology is an integral part of your study of calculus using *Calculus Concepts*. Even though the text requires technology, it does not demand any particular technology. Any graphing calculator or computer algebra system that has the functionality indicated in this *Guide* will suffice. The materials contained herein provide instruction for using the TI-83/83+ and the TI-86 in your course. A similar guide for using the TI-89 with *Calculus Concepts* is available at the Houghton Mifflin web site for the third edition of the text.

The discussions in this *Guide* are ordered to match the organization of the text chapters. These notation conventions are used to help you recognize various commands and keystrokes:

- Main keyboard keys are enclosed in rectangular boxes except for numeric keys and English alphabet letters; for example, [ENTER].

- The second function of a key is listed in parentheses after the main keyboard keystrokes used to activate the second function; for example, [2nd] [ESC] (QUIT).

- The alpha functions of keys are listed in parentheses after the main keyboard keystrokes used to activate the alpha function; for example, [ALPHA] 4 (L).

- Function keys and menu items are indicated by the main keyboard key followed by the keystroke sequence necessary to access the item and then the name of the item in brackets; for example, [F1] [Tools] 5 [Copy].

This *Guide* does not replace your calculator's instruction manual. You should refer to that manual to learn about the basic operation and use of your calculator. Calculator programs referenced in these materials (with calculator code given in the *Program Appendices* in this *Guide*) are available for download for both PC and Macintosh computers at the Houghton Mifflin web site for the third edition of *Calculus Concepts*.

I would like to thank Dave George and Jennifer King at Houghton Mifflin Company for their help and patience with this project. Any comments or suggestions concerning this *Guide* can be directed to the publisher.

This work is dedicated to my husband, James W. Reed.

Iris Fetta Reed
Clemson University

Contents

*Sections 1.2.S1 and 1.2.S2 only appear on the *Calculus Concepts* CD-ROM and web site. The instructions in these sections pertain to the *Appendix B* supplementary material for *Constructing Inverse Functions Algebraically* on pages B9–B13.

2 INGREDIENTS OF CHANGE: NONLINEAR MODELS

3 DESCRIBING CHANGE: RATES

---

*Sections 1.3.S1 through 1.3.S5 only appear on the *Calculus Concepts* CD-ROM and web site. The instructions in these sections pertain to the *Appendix B* supplementary material for *APR and APY* on pages B24–B27.

---

*Sections 8.1.S1 through 8.1.S3 only appear on the *Calculus Concepts* CD-ROM and web site. The instructions in these sections pertain to the *Appendix A: Trigonometry Basics* supplementary material on pages A2–A23.

How to Use This Guide

First, go to either part *A* or part *B* of this *Guide*, depending on which model calculator you are using. Then find the particular section in this *Guide* that references the supplement icon in the text, marked with , according to the table of contents section number. Section numbers in the contents should be read as follows:

> Section *x.y.z* of this *Guide* refers to the discussion in Chapter *x*, Section *y* of the *Calculus Concepts* text. The icon referenced is the *z*th one to appear in that particular section of the text. When a particular icon refers to more than one technology discussion, *a*, *b*, etc. follows the *Guide* section number.

For example, the detailed contents indicate that there are four different sections in this *Guide* (Sections 1.4.2a through 1.4.2d) pertaining to the second open book icon shown in Section 1.4 in *Calculus Concepts*. The discussion in Section 1.4.2c of this *Guide* is the third technology discussion pertaining to the material referenced with the second in Chapter 1, Section 1.4 of the text.

Part A

Guide for Texas Instruments TI-83 Graphing Calculator

This *Guide* is designed to offer step-by-step instruction for using your TI-83 or TI-83 Plus graphing calculator with the third edition of *Calculus Concepts: An Informal Approach to the Mathematics of Change*. A technology icon next to a particular example or discussion in the text directs you to a specific portion of this *Guide*. You should also utilize the table of contents in this *Guide* to find specific topics on which you need instruction.

Setup Instructions

Before you begin, check the TI-83 setup and make sure the settings described below are chosen. Whenever you use this *Guide*, we assume (unless instructed otherwise) that your calculator settings are as shown in Figures 1, 2, and 3.

- Press MODE and choose the settings shown in Figure 1 for the basic setup.

- Specify the statistical setup with STAT 5 [SetUpEditor] followed by 2nd 1 [,] 2nd 2 [,] 2nd 3 [,] 2nd 4 [,] 2nd 5 [,] and 2nd 6. Press ENTER to view the screen in Figure 2.

- Check the window format by pressing 2nd ZOOM (FORMAT) and choose the settings shown in Figure 3.

  - If you do not have the darkened choices shown in Figure 1 and Figure 3, use the arrow keys to move the blinking cursor over the setting you want to choose and press ENTER.

  - Press 2nd MODE (QUIT) to return to the home screen.

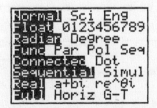

TI-83 Basic Setup

Figure 1

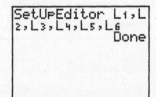

TI-83 Statistical Setup

Figure 2

TI-83 Window Setup

Figure 3

Basic Operation

You should be familiar with the basic operation of your calculator. With your calculator in hand, go through each of the following.

1. **CALCULATING**   You can type in lengthy expressions; just make sure that you use parentheses when you are not sure of the calculator's order of operations. Always enclose in parentheses any numerators and denominators of fractions and exponents that consist of more than one symbol.

Evaluate $\dfrac{1}{4*15+\frac{895}{7}}$. Enclose the denominator in parentheses so

that the addition is performed before the division into 1. It is not necessary to use parentheses around the fraction 895/7.

Evaluate $\dfrac{(-3)^4-5}{8+1.456}$. Use $\boxed{(-)}$ for the negative symbol and $\boxed{-}$ for

the subtraction sign. To clear the home screen, press $\boxed{\text{CLEAR}}$.

```
1/(4*15+895/7)
          .0053231939
((-3)^4-5)/(8+1.
456)
          8.037225042
■
```

NOTE: The numerator and denominator must be enclosed in parentheses and $-3^4 \neq (-3)^4$.

Now, evaluate $e^3*2 \approx 40.17$. The TI-83 prints the left parenthesis when you press $\boxed{\text{2nd}}$ $\boxed{\text{LN}}$ (e^x). The calculator assumes that you are inserting a right parenthesis at the end of what you type unless you insert one elsewhere. Also, if you press $\boxed{(}$ after pressing the key for e and do not include another right parenthesis, the TI-83 assumes that you are including everything following the extra parenthesis in the exponent.

```
e^(3)*2
          40.17107385
e^((3)*2
          403.4287935
e^(3*2
          403.4287935
```

2. **USING THE ANS MEMORY**   Instead of again typing an expression that was just evaluated, use the answer memory by pressing $\boxed{\text{2nd}}$ $\boxed{(-)}$ (ANS).

Calculate $\left(\dfrac{1}{4*15+\frac{895}{7}}\right)^{-1}$ using this nice shortcut.

Type Ans^{-1} by pressing $\boxed{\text{2nd}}$ $\boxed{(-)}$ (ANS) $\boxed{x^{-1}}$.

```
895/7
          127.8571429
1/(4*15+Ans)
          .0053231939
Ans⁻¹
          187.8571429
■
```

3. **ANSWER DISPLAY**   When the denominator of a fraction has no more than three digits, your calculator can provide the answer in the form of a fraction. When an answer is very large or very small, the calculator displays the result in scientific notation.

The "to a fraction" key is obtained by pressing $\boxed{\text{MATH}}$ 1 [▶Frac].

```
2/5+1/3
          .7333333333
Ans▶Frac
               11/15
.3875▶Frac
               31/80
■
```

The calculator's symbol for *times 10^{12}* is E12. Thus, 7.945E12 means 7,945,000,000,000.

The result 1.4675E⁻6 means $1.4675*10^{-6}$, which is the scientific notation expression for 0.0000014675.

4. **STORING VALUES**   Sometimes it is beneficial to store numbers or expressions for later recall. To store a number, type the number, press [STO▸] [ALPHA], type the letter corresponding to the storage location, and then press [ENTER]. To join several short commands together, use [ALPHA] [.] (:) between the statements. Note that when you join statements with a colon, only the value of the last statement is shown.

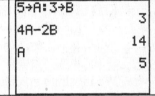

Store 5 in A and 3 in B, and then calculate $4A - 2B$.

To recall a stored value, press [ALPHA], type the letter in which the expression or value is stored, and then press [ENTER].

- There are 27 memory locations in the TI-83: A through Z and θ. Whatever you store in a particular memory location stays there until it is replaced by something else. Before you use a particular location, 0 is the stored value.

5. **ERROR MESSAGES**   When your input is incorrect, the TI-83 displays an error message.

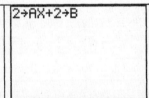 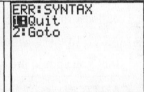

If you have more than one command on a line without the commands separated by a colon (:), an error message results when you press [ENTER].

- When you get an error message, Choose 2 [Goto] to position the cursor to where the error occurred so that you can correct the mistake or choose 1 [Quit] to begin a new line on the home screen.

- A common mistake is using the negative symbol [(−)] instead of the subtraction sign [−] or vice-versa. When you choose Goto, the TI-83 highlights the position of this error.

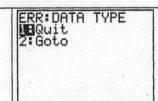

If you try to store something to a particular memory location that is being used for a different type of object, a DATA TYPE error results. Consult either *Troubleshooting the TI-83* in this *Guide* or your *TI-83 Owner's Guidebook*.

- When you are executing a program that you have transferred to or input in your calculator, you should always choose the 1: Quit option. Choosing 2: Goto will call up the program code, and you may inadvertently change the program so that it will not properly execute.

- Other error messages are referred to throughout this *Guide* at the points that they might occur in specific applications.

Chapter 1   Ingredients of Change: Functions and Linear Models

 ## 1.1 Models, Functions, and Graphs

Graphing a function in an appropriate viewing window is one of the many uses for a function that is entered in the calculator's graphing list. Because you must enter a function formula on one line (that is, you cannot write fractions and exponents the same way you do on paper), it is very important to have a good understanding of the calculator's order of operations and to use parentheses whenever they are needed.

NOTE: If you are not familiar with the basic operation of the TI-83, you should work through pages A-1 through A-3 of this *Guide* before proceeding with this material.

1.1.1a ENTERING AN EQUATION IN THE GRAPHING LIST   The graphing list contains space for 10 equations, and the output variables are called by the names Y1, Y2, ..., and Y0. When you want to graph an equation, you should enter in the graphing list. You must use X as the input variable if you intend to draw the graph of the equation or use the TI-83 table.

We illustrate graphing with the equation in Example 3 of Section 1.1: $v(t) = 3.622(1.093^t)$.

Press $\boxed{Y=}$ to access the graphing list. If there are any previously entered equations that you will no longer use, delete them from the graphing list.	To delete an equation from the Y= list, position the cursor on the line with the equation. Press $\boxed{\text{CLEAR}}$.
| For convenience, we use the first, or Y1, location in the list. We intend to graph this equation, so the input variable must be called $x$, not $t$. Enter the right-hand side of the equation, $3.622(1.093^x)$, with 3 $\boxed{.}$ 6 2 2 $\boxed{(}$ 1 $\boxed{.}$ 0 9 3 $\boxed{\wedge}$ $\boxed{X,T,\theta,n}$ $\boxed{)}$. Note that you should press $\boxed{X,T,\theta,n}$ for $x$, not the times sign key, $\boxed{\times}$. | ```
Plot1 Plot2 Plot3
\Y1∎3.622(1.093^
X)
\Y2=
\Y3=
\Y4=
\Y5=
\Y6=
``` |

CAUTION: Plot1, Plot2, and Plot3 at the top of the Y= list should *not* be darkened when you are graphing an equation and not graphing data points. If any of these is darkened, use $\boxed{\blacktriangle}$ until you are on the darkened plot name. Press $\boxed{\text{ENTER}}$ to make the name(s) not dark (that is, to *deselect* the plot). If you do not do this, you may receive a STAT PLOT error message.

1.1.1b DRAWING A GRAPH   As is the case with most applied problems in *Calculus Concepts*, the problem description indicates the valid input interval. Consider Example 3 of Section 1.1:

> The value of a piece of property between 1985 and 2005 is given by $v(t) = 3.622(1.093^t)$ thousand dollars where t is the number of years since the end of 1985.

The input interval is 1985 ($t = 0$) to 2005 ($t = 20$). Before drawing the graph of v on this interval, enter the $v(t)$ equation in the Y= list using x as the input variable. (See Section 1.1.1a of this *Guide*.) We now draw the graph of the function v for x between 0 and 20.

| Press $\boxed{\text{WINDOW}}$ to set the view for the graph. Enter 0 for Xmin and 20 for Xmax. (For 10 tick marks between 0 and 20, enter 2 for Xscl. If you want 20 tick marks, enter 1 for Xscl, etc.  Xscl does not affect the shape of the graph. Ignore the other numbers because we set their values in the next set of instructions.) | ```
WINDOW
 Xmin=0
 Xmax=20
 Xscl=2
 Ymin=-6
 Ymax=6
 Yscl=1
 Xres=1
``` |

Xmin and Xmax are, respectively, the settings of the left and right edges of the viewing screen, and Ymin and Ymax are, respectively, the settings for the lower and upper edges of the viewing screen. Xscl and Yscl set the spacing between the tick marks on the *x*- and *y*-axes. (Leave Xres set to 1 for all applications in this *Guide*.)

To have the TI-83 determine the view for the output, press ZOOM ▲ [ZoomFit]. Now press ENTER to see the graph of the function *v*.	ZOOM MEMORY 4↑ZDecimal 5:ZSquare 6:ZStandard 7:ZTrig 8:ZInteger 9:ZoomStat 0↓ZoomFit

Notice that any vertical line drawn on this graph intersects it in only one point, so the graph does represent a function.

Press WINDOW to see the view set by Zoomfit. The view has $0 \leq x \leq 20$ and $3.622 \leq y \leq 21.446...$. (Notice that ZoomFit did not change the *x*-values that you set manually.)	WINDOW Xmin=0 Xmax=20 Xscl=2 Ymin=3.622 Ymax=21.446265... Yscl=1 Xres=1

1.1.1c MANUALLY CHANGING THE VIEW OF A GRAPH We just saw how to have the TI-83 set the view for the output variable. Whenever you draw a graph, you can also manually set or change the view for the output variable. If for some reason you do not have an acceptable view of a graph or if you do not see a graph, change the view for the output variable with one of the ZOOM options or manually set the WINDOW until you see a good graph. (We will later discuss other ZOOM options.) We continue using the function *v* in Example 3 of Section 1.1, but here assume that you have not yet drawn the graph of *v*.

Press WINDOW, enter 0 for Xmin, 20 for Xmax, and (assuming we do not know what to use for the vertical view), enter some arbitrary values for Ymin and Ymax. (The graph to the right was drawn with Ymin = ‾5 and Ymax = 5). Press GRAPH.	
Evaluating Outputs on the Graphics Screen: Press TRACE. Recall that we are given in this application that the input variable is between 0 and 20. If you now type the number that you want to substitute into the function whose graph is drawn, say 0, you see the screen to the right. Note that the equation of the function whose graph you are drawing appears at the top of the screen.	Y1=3.622(1.093^X) X=0
Press ENTER and the input value is substituted in the function. The input and output values are shown at the bottom of the screen. (This method works even if you do not see any of the graph on the screen.)	Y1=3.622(1.093^X) X=0 Y=3.622
Substitute the right endpoint of the input interval into the function by pressing 20 ENTER. We see that two points on this function are approximately (0, 3.622) and (20, 21.446).	Y1=3.622(1.093^X) X=20 Y=21.446266

Press $\boxed{\text{WINDOW}}$, enter 3.5 for Ymin and 22 for Ymax, and press $\boxed{\text{GRAPH}}$. If the graph you obtain is not a good view of the function, repeat the above process using x-values in between the two endpoints to see if the output range should be extended in either direction. (Note that the choice of the values 3.5 and 22 was arbitrary. Any values close to the outputs in the points you find are also acceptable.)

NOTE: Instead of using TRACE with the exact input to evaluate outputs on the graphics screen, you could use the TI-83 TABLE or evaluate the function at 0 and 20 on the home screen. We next discuss using these features.

1.1.1d TRACING TO ESTIMATE OUTPUTS You can display the coordinates of certain points on the graph by tracing. Unlike the substitution feature of TRACE that was discussed on the previous page, the x-values that you see when tracing the graph depend on the horizontal view that you choose. The output values that are displayed at the bottom of the screen are calculated by substituting the x-values into the equation that is being graphed. We again assume that you have the function $v(x) = 3.622(1.093^x)$ entered in the Y1 location of the Y= list.

With the graph on the screen, press $\boxed{\text{TRACE}}$, press and hold $\boxed{\blacktriangleright}$ to move the trace cursor to the right, and press and hold $\boxed{\blacktriangleleft}$ to move the trace cursor to the left. The equation that you are tracing appears at the top of the graphing screen.

Trace past one edge of the screen and notice that even though you cannot see the trace cursor, the x- and y-values of points on the line are still displayed at the bottom of the screen. Also note that the graph scrolls to the left or right as you move the cursor past the edge of the current viewing screen.

Use either $\boxed{\blacktriangleright}$ or $\boxed{\blacktriangleleft}$ to move the cursor near $x \approx 15$. We estimate that y is *approximately* 13.9 when x is *about* 15.

It is important to realize that trace outputs should *never* be given as answers to a problem unless the displayed x-value is *identically* the same as the value of the input variable.

1.1.1e EVALUATING OUTPUTS ON THE HOME SCREEN The input values used in the evaluation process are *actual* values, not estimated values such as those generally obtained by tracing near a certain value. Actual values are also obtained when you evaluate outputs from the graphing screen using the process that was discussed in Section 1.1.3 of this *Guide*.

We again consider the function $v(t) = 3.622(1.093^t)$ that is in Example 3 of Section 1.1.

Using x as the input variable, enter Y1 = 3.622(1.093^X). Return to the home screen by pressing $\boxed{\text{2nd}}$ $\boxed{\text{MODE}}$ (QUIT). Go to the Y–VARS menu with $\boxed{\text{VARS}}$ $\boxed{\blacktriangleright}$ [Y–VARS]. Choose 1: Function by pressing the number 1 or $\boxed{\text{ENTER}}$.

NOTE: We choose Y1 as the function location most of the time, but you can use any of the ten available locations. If you do, replace Y1 in the instructions with the location you choose.

Choose 1: Y1 by pressing the number 1 or ENTER . Y1 appears on the home screen. Press [(], type the *x*-value to be substituted into the equation, and press [)]. Then press ENTER .	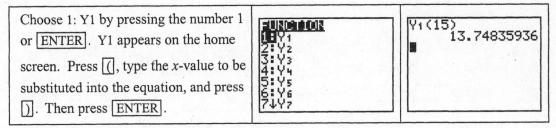	

NOTE: The *closing* parenthesis on the right is not needed if nothing else follows it.

It is now a simple matter to evaluate the function at other inputs. For instance, substitute *x* = 20 into the equation by recalling the previous entry with 2nd ENTER (ENTRY), change 15 to 20 by pressing ◄ ◄ ◄ and typing 20, and then press ENTER .	Y₁(15) 13.74835936 Y₁(20) 21.44626584
Evaluate Y1 at *x* = 0 by recalling the previous entry with 2nd ENTER (ENTRY), change 20 to 0 by pressing ◄ ◄ ◄ DEL , and then press ENTER .	Y₁(15) 13.74835936 Y₁(20) 21.44626584 Y₁(0) 3.622

1.1.1f EVALUATING OUTPUTS USING THE TABLE   Function outputs can be determined by evaluating on the graphics screen, as discussed in Section 1.1.1c, or by evaluating on the home screen as discussed in Section 1.1.1e of this *Guide*. You can also evaluate functions using the TI-83 TABLE. When you use the table, you can either enter specific input values and find the outputs or generate a list of input and output values in which the inputs begin with TblStart and differ by ΔTbl.

Let's use the TABLE to evaluate the function $v(t) = 3.622(1.093^t)$ at the input $t = 15$. Even though you can use any of the function locations, we again choose to use Y1. Press Y= , clear the function locations, and enter 3.622(1.093^X) in location Y1 of the Y= list.

Choose TABLE SETUP with 2nd WINDOW (TBLSET). To generate a list of values beginning with 13 such that the table values differ by 1, enter 13 in the TblStart location, 1 in the ΔTbl location, and choose AUTO in the Indpnt: and Depend: locations. (Remember that you "choose" a certain setting by moving the cursor to that location and pressing ENTER .	TABLE SETUP TblStart=13 ΔTbl=1 Indpnt: **AUTO** Ask Depend: **AUTO** Ask
Press 2nd GRAPH (TABLE), and observe the list of input and output values. Notice that you can scroll through the table with ▼ , ▲ , ◄ , and/or ► . The table values may be rounded in the table display. You can see more of the output by highlighting a particular value and viewing the bottom of the screen.	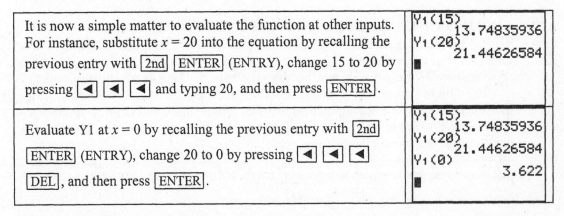

If you are interested in evaluating a function at inputs that are not evenly spaced and/or you only need a few outputs, you should use the ASK feature of the table instead of using AUTO. Note that when using ASK, the settings for TblStart and ΔTbl do not matter.

Return to the table set-up with $\boxed{\text{2nd}}$ $\boxed{\text{WINDOW}}$ (TBLSET), and choose ASK in the Indpnt: location. Press $\boxed{\text{ENTER}}$. Press $\boxed{\text{2nd}}$ $\boxed{\text{GRAPH}}$ (TABLE) to go to the table and enter the x-values.	```
TABLE SETUP
 TblStart=13
 △Tbl=1
Indpnt: Auto Ask
Depend: Auto Ask
``` ```
 X   Y1
15 13.748
 0   3.622
20 21.446

Y1=13.7483593553
``` |

NOTE: Unwanted entries or values from a previous problem can be cleared with $\boxed{\text{DEL}}$.

1.1.1g FINDING INPUT VALUES USING THE SOLVER Your calculator solves for the input values of any equation that is in the form *"expression = 0"*. This means that all terms must be on one side of the equation and 0 must be on the other side before you enter the equation into the calculator. The expression can, but does not have to, use x as the input variable.

The TI-83 offers several methods of solving for input variables. We first illustrate using the SOLVER. (Solving using graphical methods will be discussed after using the SOLVER is explored.) You can refer to an equation that you have already entered in the Y= list or you can enter the equation in the solver.

| Return to the home screen with $\boxed{\text{2nd}}$ $\boxed{\text{MODE}}$ (QUIT). Access the solver by pressing $\boxed{\text{MATH}}$ 0. If there are no equations stored in the solver, you will see the screen displayed on the right – or | ```
EQUATION SOLVER
eqn:0=■
``` |
| – if the solver has been previously used, you will see a screen similar to the one shown on the right. If this is the case, press $\boxed{\blacktriangle}$ and $\boxed{\text{CLEAR}}$ to delete the old equation. You should then have the screen shown in the previous step. | ```
2X²-3=0
 X=2
 bound={-1E99,1…
``` |

Let's now use the solver to answer the question in part *e* of Example 3 in Section 1.1: "When did the land value reach $20,000?" Because the land value is given by $v(t) = 3.622(1.093^t)$ thousand dollars where t is the number of years after the end of 1985, we are asked to *solve* the equation $3.622(1.093^t) = 20$. That is, we are asked to find the input value t that makes this equation a true statement. We must enter the equation into the solver with a 0 on one side, so subtract 20 from each side of $3.622(1.093^t) = 20$ to obtain the equation $3.622(1.093^t) - 20 = 0$.

| If you already have Y1 = 3.622(1.093^X) in the graphing list, you can refer to the function as Y1 in the SOLVER. If not, you can enter 3.622(1.093^X) instead of Y1 in the eqn: location of the SOLVER. Press $\boxed{\text{ENTER}}$. | ```
EQUATION SOLVER
eqn:0=Y1-20
``` |
| If you need to edit the equation, press $\boxed{\blacktriangle}$ until the previous screen reappears. Edit the equation and then return here. With the cursor in the X location, enter a guess – say 19. (You could have also used as a guess* the value that was in the X location when you accessed the SOLVER.) | ```
Y1-20=0
 X=19■
 bound={-1E99,1…
``` |

*More information on entering a guess appears at the end of this discussion.

CAUTION: You should not change anything in the "bound" location of the SOLVER. The values in that location are the ones between which the TI-83 searches[1] for a solution. If you should accidentally change anything in this location, exit the solver, and begin the entire process again. (The bound is automatically reset when you exit the SOLVER.)

Be certain that the cursor is on the line corresponding to the input variable for which you are solving (in this example, X). Solve for the input by pressing ALPHA ENTER (SOLVE). The answer to the original question is that the land value was $20,000 about 19.2 years after 1985 – *i.e.*, in the year 2005.	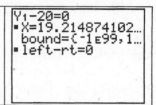

- Notice the black dot that appears next to X and next to the last line on the above screen. This is the TI-83's way of telling you that a solution has been found. The bottom line on the screen that states left – rt = 0 indicates the value found for X is an exact solution since both sides of the equation evaluate to the same quantity.

- If a solution continues beyond the edge of the calculator screen, you see "..." to the right of the value. Be certain that you press and hold ▶ to scroll to the end of the number.

  The value may be given in scientific notation, and the portion that you cannot see determines the location of the decimal point. (See *Basic Operation, #3*, in this *Guide*.)

1.1.1h HOW TO DETERMINE A GUESS TO USE IN THE EQUATION SOLVER What you use in the solver as a guess tells the TI-83 where to start looking for the answer. How close your guess is to the actual answer is not very important unless there is more than one solution to the equation. *If the equation has more than one answer, the solver will return the solution that is closest to the guess you supply.* In such cases, you need to know how many answers you should search for and their approximate locations.

Three of the methods that you can use to estimate the value of a guess for an answer from the SOLVER follow. We illustrate these methods using the land value function from Example 3 of Section 1.1 and the equation $v(t) = 3.622(1.093^t) = 20$.

1. Enter the function in some location of the graphing list – say Y1 = 3.622(1.093^X) and draw a graph of the function. Press TRACE and hold down either ▶ or ◀ until you have an estimate of where the output is 20. Use this X-value, 19 or 19.4 or 19.36, as your guess in the SOLVER.	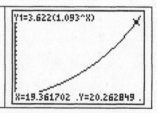

Remember that you can use any letter to represent the input in the solver, but to graph a function, you must use X as the input variable.

2. Enter the left- and right-hand sides of the equation in two different locations of the Y= list – say Y1 = 3.622(1.093^X) and Y2 = 20. With the graph on the screen, press TRACE and hold down either ▶ or ◀ until you get an estimate of the X-value where the curve crosses the horizontal line representing 20.	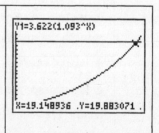

---

[1] It is possible to change the bound if the calculator has trouble finding a solution to a particular equation. This, however, should happen rarely. Refer to the *TI-83 Graphing Calculator Guidebook* for details.

<table>
<tr><td>

3. Use the AUTO setting in the TABLE, and with [▲] or [▼] scroll through the table until a value near the estimated output is found. Use this X-value or a number near it as your guess in the SOLVER. (Refer to Section 1.1.1f of this *Guide* to review the instructions for using the TABLE.)

</td><td>

X	Y₁	Y₂
18	17.952	20
19	19.621	20
20	21.446	20
21	23.441	20
22	25.621	20
23	28.003	20
24	30.608	20

Y₁=19.621469201

</td></tr>
</table>

1.1.1i GRAPHICALLY FINDING INTERCEPTS Finding the input value at which the graph of a function crosses the vertical and/or horizontal axis can be found graphically or by using the SOLVER. Remember the process by which we find intercepts:

- To find the y-intercept of a function $y = f(x)$, set $x = 0$ and solve the resulting equation.

- To find the x-intercept of a function $y = f(x)$, set $y = 0$ and solve the resulting equation.

An *intercept* is the where the graph crosses or touches an axis. Also remember that the x-intercept of the function $y = f(x)$ has the same value as the root or solution of the equation $f(x) = 0$. **Thus, finding the x-intercept of the graph of $f(x) - c = 0$ is the same as solving the equation $f(x) = c$.**

We illustrate this method with a problem similar to the one in Activity 36 in Section 1.1 of *Calculus Concepts*. You should practice both methods by solving $3.622(1.093^x) = 20$ using the following graphical methods and by solving the equation below by using the SOLVER.

Suppose we are asked to find the input value of $f(x) = 3x - 0.8x^2 + 4$ that corresponds to the output $f(x) = 2.3$. That is, we are asked to find x such that $3x - 0.8x^2 + 4 = 2.3$. Because this function is not given in a context, we have no indication of an interval of input values to use when drawing the graph. We will use the ZOOM features to set an initial view and then manually set the WINDOW until we see a graph that shows the important points of the function (in this case, the intercept or intercepts.) You can solve this equation graphically using either the *x-intercept method* or the *intersection method*. We present both, and you should use the one you prefer.

X-INTERCEPT METHOD for solving the equation $f(x) - c = 0$:

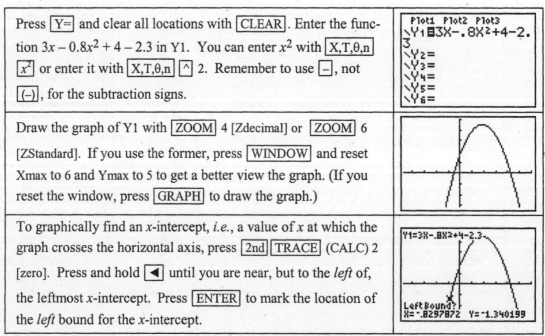

<table>
<tr><td>

Press [Y=] and clear all locations with [CLEAR]. Enter the function $3x - 0.8x^2 + 4 - 2.3$ in Y1. You can enter x^2 with [X,T,θ,n] [x^2] or enter it with [X,T,θ,n] [^] 2. Remember to use [−], not [(−)], for the subtraction signs.

</td><td>

Plot1 Plot2 Plot3
\Y1☐3X-.8X²+4-2.
3
\Y2=
\Y3=
\Y4=
\Y5=
\Y6=

</td></tr>
<tr><td>

Draw the graph of Y1 with [ZOOM] 4 [Zdecimal] or [ZOOM] 6 [ZStandard]. If you use the former, press [WINDOW] and reset Xmax to 6 and Ymax to 5 to get a better view the graph. (If you reset the window, press [GRAPH] to draw the graph.)

</td><td>

</td></tr>
<tr><td>

To graphically find an x-intercept, *i.e.*, a value of x at which the graph crosses the horizontal axis, press [2nd] [TRACE] (CALC) 2 [zero]. Press and hold [◄] until you are near, but to the *left* of, the leftmost x-intercept. Press [ENTER] to mark the location of the *left* bound for the x-intercept.

</td><td>

Y1=3X-.8X²+4-2.3

Left Bound?
X=-.8297872 Y=-1.340199

</td></tr>
</table>

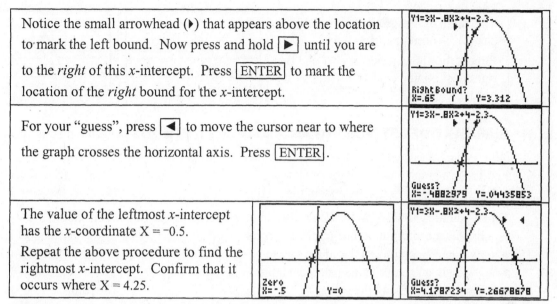

Notice the small arrowhead (▶) that appears above the location to mark the left bound. Now press and hold ▶ until you are to the *right* of this *x*-intercept. Press ENTER to mark the location of the *right* bound for the *x*-intercept.	Y1=3X-.8X2+4-2.3 Right Bound? X=.65 Y=3.312
For your "guess", press ◀ to move the cursor near to where the graph crosses the horizontal axis. Press ENTER.	Y1=3X-.8X2+4-2.3 Guess? X=-.4882979 Y=.04435853
The value of the leftmost *x*-intercept has the *x*-coordinate X = −0.5. Repeat the above procedure to find the rightmost *x*-intercept. Confirm that it occurs where X = 4.25.	Zero X=-.5 Y=0
	Y1=3X-.8X2+4-2.3 Guess? X=4.1787234 Y=.26678678

NOTE: If this process does not return the correct value for the intercept that you are trying to find, you have probably not included the place where the graph crosses the axis between the two bounds (*i.e.,* between the ▶ and ◀ marks on the graph.)

INTERSECTION METHOD for solving the equation $f(x) = c$:

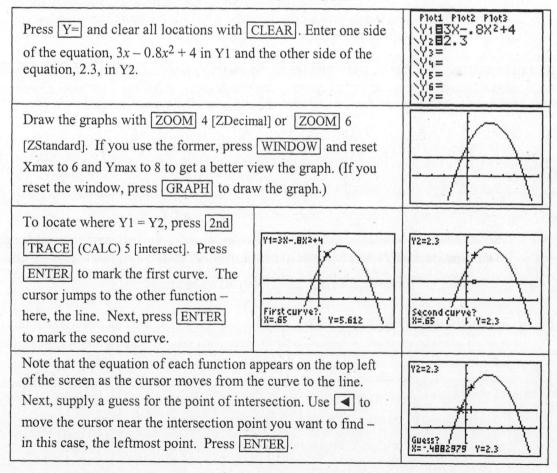

Press Y= and clear all locations with CLEAR. Enter one side of the equation, $3x − 0.8x^2 + 4$ in Y1 and the other side of the equation, 2.3, in Y2.	Plot1 Plot2 Plot3 \Y1■3X-.8X2+4 \Y2■2.3 \Y3= \Y4= \Y5= \Y6= \Y7=
Draw the graphs with ZOOM 4 [ZDecimal] or ZOOM 6 [ZStandard]. If you use the former, press WINDOW and reset Xmax to 6 and Ymax to 8 to get a better view the graph. (If you reset the window, press GRAPH to draw the graph.)	
To locate where Y1 = Y2, press 2nd TRACE (CALC) 5 [intersect]. Press ENTER to mark the first curve. The cursor jumps to the other function – here, the line. Next, press ENTER to mark the second curve.	Y1=3X-.8X2+4 First curve? X=.65 Y=5.612
	Y2=2.3 Second curve? X=.65 Y=2.3
Note that the equation of each function appears on the top left of the screen as the cursor moves from the curve to the line. Next, supply a guess for the point of intersection. Use ◀ to move the cursor near the intersection point you want to find – in this case, the leftmost point. Press ENTER.	Y2=2.3 Guess? X=-.4882979 Y=2.3

The value of the leftmost x-intercept has the x-coordinate X = $^-$0.5. Repeat the above procedure to find the rightmost x-intercept. Confirm that it occurs where X = 4.25.	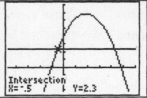	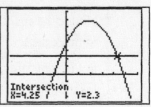

1.1.1j SUMMARY OF ESTIMATING AND SOLVING METHODS Use the method you prefer.

When you are asked to *estimate* or *approximate* an output or an input value, you can:

- Trace a graph entered in the Y= list   (Section 1.1.1d)
- Use close values obtained from the TABLE   (Section 1.1.1f)

When you are asked to *find* or *determine* an output or an input value, you should:

- Evaluate an output on the graphics screen   (Section 1.1.1c)
- Evaluate an output on the home screen   (Section 1.1.1e)
- Evaluate an output value using the table   (Section 1.1.1f)
- Find an input using the solver   (Sections 1.1.1g, 1.1.1h)
- Find an input value from the graphic screen (using the x-intercept
  method or the intersection method)   (Section 1.1.1i)

1.2 Constructed Functions

Your calculator can find output values of and graph combinations of functions in the same way that you do these things for a single function. The only additional information you need is how to enter constructed functions in the graphing list.

1.2.1a FINDING THE SUM, DIFFERENCE, PRODUCT, QUOTIENT, OR COMPOSITE FUNCTION Suppose that a function f has been entered in Y1 and a function g has been entered in Y2. Your calculator will evaluate and graph these constructed functions:

Enter Y1 + Y2 in Y3 to obtain the sum function $(f + g)(x) = f(x) + g(x)$.
Enter Y1 − Y2 in Y4 to obtain the difference function $(f - g)(x) = f(x) - g(x)$.
Enter Y1*Y2 in Y5 to obtain the product function $(f \cdot g)(x) = f(x) \cdot g(x)$.

Enter Y1/Y2 in Y6 to obtain the quotient function $(f \div g)(x) = \dfrac{f(x)}{g(x)}$.

Enter Y1(Y2) in Y7 to obtain the composite function $(f \circ g)(x) = f(g(x))$.

The functions can be entered in any location in the Y= list. Although the TI-83 will not give an algebraic formula for a constructed function, you can check your final answer by evaluating your constructed function and the calculator-constructed function at several different points to see if they yield the same outputs.

1.2.1b FINDING A DIFFERENCE FUNCTION We illustrate this technique with the functions that are given on page 19 in Section 1.2 of *Calculus Concepts*: sales = $S(t) = 3.570(1.105^t)$ million dollars and costs = $C(t) = -39.2t^2 + 540.1t + 1061.0$ thousand dollars t years after 1996.

Press Y= . Clear previously entered equations with CLEAR . Enter S in Y1 by pressing 3 . 570 (1 . 105 ^ X,T,θ,n) ENTER and enter C in Y2 by pressing (−) 39 . 2 X,T,θ,n x^2 + 540 . 1 X,T,θ,n + 1061 ENTER .	

The difference function, the profit $P(x) = S(x) - 0.001C(x) =$ Y1 − 0.001Y2, is entered in Y3 with $\boxed{\text{VARS}}$ $\boxed{\blacktriangleright}$ 1 [Function] 1 [Y1] $\boxed{-}$ $\boxed{.}$ 001 $\boxed{\text{VARS}}$ $\boxed{\blacktriangleright}$ 1 [Function] 2 [Y2].	Plot1 Plot2 Plot3 \Y1■3.570(1.105^X) \Y2■-39.2X²+540.1X+1061 \Y3■Y1-.001Y2 \Y4= \Y5=
To find the profit in 1998, evaluate Y3 when X = 2. Return to the home screen with $\boxed{\text{2nd}}$ $\boxed{\text{MODE}}$ (QUIT). Press $\boxed{\text{VARS}}$ $\boxed{\blacktriangleright}$ 1 [Function] 3 [Y3] $\boxed{(}$ 2 $\boxed{)}$ $\boxed{\text{ENTER}}$ to see the result. We find that the profit in 1998 was $P(2) \approx 2.375$ million dollars.	Y3(2) 2.37465925

- You can evaluate on the home screen, graphics screen, or in the table. We choose to use the home screen, but you should choose the method you prefer.

1.2.1c FINDING A PRODUCT FUNCTION We illustrate this technique with the functions given on page 21 of Section 1.2 of *Calculus Concepts*: Milk price = $M(x) = 0.007x + 1.492$ dollars per gallon on the xth day of last month and milk sales = $G(x) = 31 - 6.332(0.921^x)$ gallons of milk sold on the xth day of last month.

Press $\boxed{\text{Y=}}$, and clear each previously entered equation with $\boxed{\text{CLEAR}}$. Enter M in Y1 by pressing $\boxed{.}$ 007 $\boxed{\text{X,T,θ,n}}$ $\boxed{+}$ 1 $\boxed{.}$ 492 $\boxed{\text{ENTER}}$ and input G in Y2 by pressing 31 $\boxed{-}$ 6 $\boxed{.}$ 332 $\boxed{(}$ $\boxed{.}$ 921 $\boxed{\wedge}$ $\boxed{\text{X,T,θ,n}}$ $\boxed{)}$ $\boxed{\text{ENTER}}$.	Plot1 Plot2 Plot3 \Y1■.007X+1.492 \Y2■31-6.332(.921^X) \Y3=■ \Y4= \Y5= \Y6=
Enter the product function $T(x) = M(x) \cdot G(x) =$ Y1·Y2 in Y3 with the keystrokes $\boxed{\text{VARS}}$ $\boxed{\blacktriangleright}$ 1 [Function] 1 [Y1] $\boxed{\text{VARS}}$ $\boxed{\blacktriangleright}$ 1 [Function] 2 [Y2].	Plot1 Plot2 Plot3 \Y1■.007X+1.492 \Y2■31-6.332(.921^X) \Y3■Y1Y2 \Y4= \Y5= \Y6=

NOTE: You do not have to, but you can, use $\boxed{\times}$ between Y1 and Y2 to indicate a product function. You *cannot* use parentheses to indicate the product function because the TI-83 will think that you are entering Y3 as a composite function.

To find milk sales on the 5th day of last month, evaluate Y3 at X = 5. We choose to do this on the home screen. Return to the home screen with $\boxed{\text{2nd}}$ $\boxed{\text{MODE}}$ (QUIT). Press $\boxed{\text{VARS}}$ $\boxed{\blacktriangleright}$ 1 [Function] 3 [Y3] $\boxed{(}$ 5 $\boxed{)}$ $\boxed{\text{ENTER}}$ to see the result. We find that milk sales were $T(5) \approx \$40.93$.	Y3(5) 40.92965522 ■

1.2.2a CHECKING YOUR ANSWER FOR A COMPOSITE FUNCTION We illustrate this technique with the functions that are given on page 23 of Section 1.2 of *Calculus Concepts*: altitude = $F(t) = -222.22t^3 + 1755.95t^2 + 1680.56t + 4416.67$ feet above sea level where t is the time into flight in minutes and air temperature = $A(F) = 277.897(0.99984^F) - 66$ degrees Fahrenheit where F is the number of feet above sea level. Remember that when you enter functions in the Y= list, you must use X as the input variable.

Press [Y=], and clear each previously entered equation with [CLEAR]. Enter F in Y1 by pressing [(-)] 222 [.] 22 [X,T,θ,n] [^] 3 [+] 1755 [.] 95 [X,T,θ,n] [x^2] [+] 1680 [.] 56 [X,T,θ,n] [+] 4416 [.] 67 [ENTER] and enter A in Y2 by pressing 277 [.] 897 [(] [.] 99984 [^] [X,T,θ,n] [)] − 66 [ENTER].

```
Plot1 Plot2 Plot3
\Y1=-222.22X^3+1
755.95X²+1680.56
X+4416.67
\Y2=277.897(.999
84^X)-66
\Y3=
\Y4=
```

Enter the composite function $(A \circ F)(x) = A(F(x)) = Y2(Y1)$ in Y3 with the keystrokes [VARS] [▶] 1 [Function] 2 [Y2] [(] [VARS] [▶] 1 [Function] 1 [Y1] [)] [ENTER].

```
Plot1 Plot2 Plot3
\Y1=-222.22X^3+1
755.95X²+1680.5
6X+4416.67
\Y2=277.897(.999
84^X)-66
\Y3=Y2(Y1)
\Y4=
```

Next, enter your algebraic answer for the composite function in Y4. Be certain that you enclose the exponent in Y4 (the function in Y1) in parentheses!

(The composite function in the text is the one that appears to the right, but you should enter the function that you have found for the composite function.)

```
Plot1 Plot2 Plot3
\Y2=277.897(.999
84^X)-66
\Y3=Y2(Y1)
\Y4=277.897(.999
84^(-222.22X^3+1
755.95X²+1680.56
X+4416.67))-66
```

We now wish to check that the algebraic answer for the composite function is the same as the calculator's composite function by evaluating both functions at several different input values. You can do these evaluations on the home screen, but as seen below, using the table is more convenient.

1.2.2b TURNING FUNCTIONS OFF AND ON IN THE GRAPHING LIST Note in the prior illustration that we are interested in the output values for only Y3 and Y4. However, the table will list values for all functions that are turned on. (A function is *turned on* when the equals sign in its location in the graphing list is darkened.) We now wish to turn off Y1 and Y2.

Press [Y=] and, using one of the arrow keys, move the cursor until it covers the darkened = in Y1. Press [ENTER]. Then, press [▼] until the cursor covers the darkened = in Y2. Press [ENTER]. Y1 and Y2 are now turned off.

```
Plot1 Plot2 Plot3
\Y1=-222.22X^3+1
755.95X^2+1680.5
6X+4416.67
\Y2=277.897(.999
84^X)-66
\Y3=Y2(Y1)
\Y4=277.897(.999
```

A function is *turned off* when the equals sign in its location in the graphing list is not dark. To turn a function back on, simply reverse the above process to make the equal sign for the function dark. When you draw a graph, the TI-83 graphs of all functions that are turned on. You may at times wish to keep certain functions entered in the graphing list but not have them graph and not have their values shown in the table. Such is the case in this illustration.

We now return to checking to see that Y3 and Y4 represent the same function.

Choose the ASK setting in the table setup so that you can check several different values for both Y3 and Y4. Recall that you access the table setup with [2nd] [WINDOW] (TBLSET). Move the cursor to ASK in the Indpnt: location and press [ENTER].

```
TABLE SETUP
 TblStart=13
 ΔTbl=1
Indpnt: Auto ASK
Depend:  Auto Ask
```

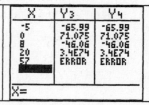

Press [2nd] [GRAPH] (TABLE), type in the *x*-value(s) at which the function is to be evaluated, and press [ENTER] after each one. We see that because all these outputs are the same for each function, you can be fairly sure that your answer is correct.

Why does ERROR appear in the table when $x = 57$? Look at the value when $x = 20$; it is very large! The computational limits of the calculator have been exceeded when $x = 57$. The TI-83 calls this an OVERFLOW error.

1.2.3 GRAPHING A PIECEWISE CONTINUOUS FUNCTION

Piecewise continuous functions are used throughout the text. You will need to use your calculator to graph and evaluate outputs of piecewise continuous functions. Several methods can be used to draw the graph of a piecewise function. One of these is presented below using the function that appears in Example 2 of Section 1.2 in *Calculus Concepts*:

The population of West Virginia from 1985 through 1999 can be modeled by

$$P(t) = \begin{cases} -23.373t + 3892.220 \text{ thousand people} & \text{when } 85 \le t < 90 \\ -1.013t^2 + 193.164t - 7387.836 \text{ thousand people} & \text{when } 90 \le t \le 99 \end{cases}$$

where *t* is the number of years since 1900.

Clear any functions that are in the Y= list. Using X as the input variable, enter each piece of the function in a separate location. We use locations Y1 and Y2.

Next, we form the formula for the piecewise function in Y3.

Parentheses <u>must</u> be used around the function portions and the inequality statements that tell the calculator which side of the break point to graph each part of the piecewise function.

With the cursor inY3, press [(] [VARS] [▶] 1 [Function] 1 [Y1] [)] [(] [X,T,θ,n] [2nd] [MATH] [TEST] 5 [<] 90 [)] [+] [(] [VARS] [▶] 1 [Function] 2 [Y2] [)] [(] [X,T,θ,n] [2nd] [MATH] [TEST] 4 [≥] 90 [)].

Your calculator draws graphs by connecting function outputs wherever the function is defined. However, this function breaks at $x = 90$. The TI-83 will connect the two pieces of *P* unless you tell it not to do so. Whenever you draw graphs of piecewise functions, set your calculator to Dot mode as described below so that it will not connect the function pieces at the break point.

Turn off Y1 and Y2. Place the cursor on the first line of Y3 and press [◀] until the blinking cursor is on the slanted line[2] to the *left* of Y3. Press [ENTER] 6 times. You should see the solid slanted line change shapes until it becomes a dotted slanted line (as shown to the right). Press [▶] several times to return the cursor to the right of the equals sign.

---

[2] The different graph styles you can draw from this location are described in more detail on pages 3-9 through 3-10 in your *TI-83 Owner's Guidebook.*

NOTE: The method described above places individual functions in Dot mode. The functions return to standard (Connected) mode when the function locations are cleared. If you want to put all functions in Dot mode at the same time, press MODE ▼ ▼ ▼ ▼ ▶ [Dot] ENTER . However, if you choose to set Dot mode in this manner, you must return to the MODE screen, select Connected, and press ENTER to take the TI-83 out of Dot mode.

Now, set the window. The function *P* is defined only when the input is between 85 and 99. So, we evaluate *P*(85), *P*(90), and *P*(99) to help when setting the vertical view.	Y₃(85)
	              1905.515
	Y₃(90)
	              1791.624
	Y₃(99)
	              1806.987
	▪

Note that if you attempt to set the window using ZoomFit as described in Section 1.1.1b of this *Guide*, the picture is not very good and you will probably want to manually reset the height of the window as described below.

We set the lower and upper endpoints of the input interval as Xmin and Xmax, respectively. Press WINDOW , set Xmin = 85, Xmax = 99, Ymin ≈ 1780, and Ymax ≈ 1910. Press GRAPH .	
Reset the window if you want a closer look at the function around the break point. The graph to the right was drawn using Xmin = 89, Xmax = 91, Ymin = 1780, and Ymax = 1810. (The visible breaks in the left piece of the graph occur because we are graphing in dot mode, not connected mode.)	

You can find function values by evaluating outputs on the home screen or using the table. Either evaluate Y3 or carefully look at the inequalities in the function *P* to determine whether Y1 or Y2 should be evaluated to obtain each particular output.	Y₃(87)
	              1858.769
	Y₁(87)
	              1858.769
	Y₃(98)
	              1813.384
	Y₂(98)▪

1.3 Limits: Functions, Limits, and Continuity

The TI-83 table is an essential tool when you estimate end behavior numerically. Even though rounded values are shown in the table due to space limitations, the TI-83 displays at the bottom of the screen many more decimal places for a particular output when you highlight that output.

1.3.1a NUMERICALLY ESTIMATING END BEHAVIOR   Whenever you use the TI-83 to estimate end behavior, set the TABLE to ASK mode. We illustrate using the function *u* that appears in Example 1 of Section 1.3 in *Calculus Concepts*:

Press Y= and use CLEAR to delete all previously entered functions. Enter $u(x) = \dfrac{3x^2 + x}{10x^2 + 3x + 2}$. Be certain to enclose the numerator and denominator of the fraction in parentheses.	Plot1 Plot2 Plot3 \Y₁◼(3X^2+X)/(10 X²+3X+2) \Y₂= \Y₃= \Y₄= \Y₅= \Y₆=

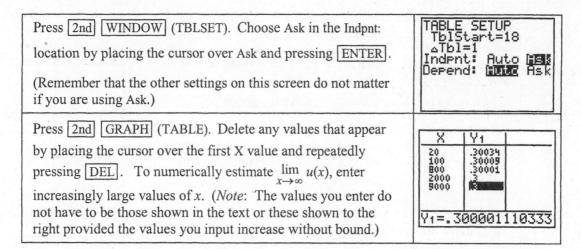

	TABLE SETUP
Press 2nd WINDOW (TBLSET). Choose Ask in the Indpnt: location by placing the cursor over Ask and pressing ENTER. (Remember that the other settings on this screen do not matter if you are using Ask.)	TblStart=18 △Tbl=1 Indpnt: Auto **Ask** Depend: **Auto** Ask
Press 2nd GRAPH (TABLE). Delete any values that appear by placing the cursor over the first X value and repeatedly pressing DEL. To numerically estimate $\lim\limits_{x \to \infty} u(x)$, enter increasingly large values of x. (*Note*: The values you enter do not have to be those shown in the text or these shown to the right provided the values you input increase without bound.)	X | Y1 20 | .30034 100 | .30009 800 | .30001 2000 | .3 9000 | **.3** Y1=.300001110333

CAUTION: Your instructor will very likely have you write the table you construct on paper. Be certain to highlight the rounded values in the output column of the table and look on the bottom of the screen to see what these values actually are.

ROUNDING OFF: Recall that *rounded off* (also called *rounded* in this *Guide*) means that if one digit past the digit of interest is less than 5, other digits past the digit of interest are dropped. If one digit past the one of interest is 5 or over, the digit of interest is increased by 1 and the remaining digits are dropped.

RULE OF THUMB FOR DETERMINING LIMITS FROM TABLES: Suppose that you are asked to give $\lim\limits_{x \to \infty} u(x)$ accurate to 3 decimal places. Observe the Y1 values in the table until you see that the output is the same value to one more decimal place (here, to 4 decimal places) for 3 consecutive outputs. Then, round the last of the repeated values off to the requested 3 places for the desired limit. Your instructor may establish a different rule from this one, so be sure to ask.

Using this Rule of Thumb and the results that are shown on the last calculator screen, we estimate that, to 3 decimal places, $\lim\limits_{x \to \infty} u(x) = 0.300$. We now need to estimate $\lim\limits_{x \to -\infty} u(x)$.

	X | Y1
Delete the values currently in the table with DEL. To estimate the negative end behavior of u, enter negative values with increasingly large magnitudes. (*Note*: Again, the values that you enter do not have to be those shown in the text or these shown to the right.)	-50 | .29977 -100 | .29989 -500 | .29998 -1500 | .29999 -9500 | **.3** Y1=.29999894667

Because the output 0.2999... appears three times in a row, we estimate $\lim\limits_{x \to -\infty} u(x) = 0.300$.

CAUTION: It is not the final value, but a sequence of several values, that is important when determining limits. If you enter a very large or very small value, you may exceed the limits of the TI-83's capability and obtain an incorrect number. Always look at the sequence of values obtained to make sure that all values that are found make sense.

1.3.1b GRAPHICALLY ESTIMATING END BEHAVIOR The graph of the function u in Example 1 of Section 1.3 in *Calculus Concepts* can be used to confirm our numerically estimated end behavior. Even though the ZOOM menu of the TI-83 can be used with this process for some functions, the graph of u is lost if you use Option 3 in the ZOOM menu (Zoom Out). We therefore manually set the window to zoom out on the horizontal axis.

Have $u(x) = \dfrac{3x^2 + x}{10x^2 + 3x + 2}$ in the Y1 location of the Y= list. (Be certain that you remember to enclose both the numerator and denominator of the fraction in parentheses.) A graph drawn with ZOOM 4 [ZDecimal] is a starting point.	
We estimated the limit as x gets very large or very small to be 0.3. Now, $u(0) = 0$, and it does appears from the graph that u is never negative. To obtain a better view of the outputs, set a WINDOW with values such as Xmin = −10, Xmax = 10, Ymin = 0, and Ymax = 0.35. Press GRAPH .	
To examine the limit as x gets larger and larger (*i.e.,* to *zoom out* on the positive x-axis), change the WINDOW so that Xmax = 100, view the graph with GRAPH , change the WINDOW so that Xmax = 1000, view the graph with GRAPH , and so forth. Press TRACE and hold down ▶ on each graph screen to view the outputs. The output values confirm our numerical estimate.	Y1=(3X²+X)/(10X²+3X+2) X=7337.766 ▬ Y=.30000136 ▪
Repeat the process as x gets smaller and smaller, but change Xmin rather than Xmax after drawing each graph. The graph to the right was drawn with Xmin = −10,000, Xmax = 10, Ymin = 0, and Ymax = 0.35. Press TRACE and hold ◀ while on each graph screen to view some of the outputs and confirm the numerical estimates.	Y1=(3X²+X)/(10X²+3X+2) X=-8083.191 ▪ Y=.29999876

1.3.2a NUMERICALLY ESTIMATING THE LIMIT AT A POINT Whenever you numerically estimate the limit at a point, you should again set the TABLE to ASK mode. We illustrate using the function u that appears in Example 2 of Section 1.3 in *Calculus Concepts*:

Enter $u(x) = \dfrac{3x}{9x + 2}$ in some Y= list location, say Y1. Have TBLSET set to Ask, and press 2nd GRAPH (TABLE) to return to the table.	Plot1 Plot2 Plot3 \Y1◼3X/(9X+2) \Y2= \Y3= \Y4= \Y5= \Y6= \Y7=
Delete the values currently in the table with DEL . To numerically estimate $\lim\limits_{x \to ^-2/9^-} u(x)$, enter values to the *left* of, and *closer and closer* to, $-2/9 = -0.222222\ldots$. Because the output values appear to become larger and larger, we estimate that the limit does not exist and write $\lim\limits_{x \to ^-2/9^-} u(x) \to \infty$.	X | Y1 -.23 | 9.8571 -.223 | 95.571 -.2223 | 952.71 -.2222 | 9524.1 ◼◼◼◼◼ | 95238 X=-.222223
Delete the values currently in the table. To numerically estimate $\lim\limits_{x \to ^-2/9^+} u(x)$, enter values to the *right* of, and *closer and closer* to, $-2/9$. Because the output values appear to become larger and larger, we estimate that $\lim\limits_{x \to ^-2/9^+} u(x) \to -\infty$.	X | Y1 -.21 | -5.727 -.218 | -17.21 -.2218 | -175.1 -.2222 | -1754 ◼◼◼◼◼ | -17544 X=-.222218

1.3.2b GRAPHICALLY ESTIMATING THE LIMIT AT A POINT   A graph can be used to estimate a limit at a point or to confirm a limit that you estimate numerically. The procedure usually involves *zooming in* on a graph to confirm that the limit at a point exists or *zooming out* to validate that a limit does not exist. We again illustrate using the function *u* that appears in Example 2 of Section 1.3 in *Calculus Concepts*.

Have the function $u(x) = \dfrac{3x}{9x + 2}$ entered in some location of the Y= list, say Y1. A graph drawn with ZOOM 4 [ZDecimal] or ZOOM 6 [ZStandard] is not very helpful.	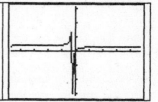

To confirm that $\displaystyle\lim_{x \to -2/9^-} u(x)$ and $\displaystyle\lim_{x \to -2/9^+} u(x)$ do not exist, we are interested in values of *x* that are near to and on either side of $-2/9$.

Choose input values close to $-0.222222...$ for the *x*-view and experiment with different *y* values until you find an appropriate vertical view. Use these values to manually set a window such as that shown to the right. Draw the graph with GRAPH .	WINDOW Xmin=-.42 Xmax=-.02 Xscl=1 Ymin=-7 Ymax=7 Yscl=1 Xres=1
The vertical line appears because the TI-83 is set to Connected mode. Place the TI-83 in Dot mode or place the function Y1 in the Y= list in Dot mode (see page A16) and redraw the graph.	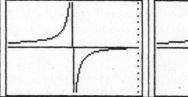

It appears from this graph that as *x* approaches $-2/9$ from the left that the output values increase without bound and that as *x* approaches $-2/9$ from the right that the output values decrease without bound. Choosing smaller Ymin values and larger Ymax values in the Window and tracing the graph as *x* approaches $-2/9$ from either side confirms this result.

Graphically Estimating a Limit at a Point when the Limit Exists: The previous illustrations involved zooming on a graph by manually setting the window. You can also zoom with the ZOOM menu of the calculator. We next describe this method by *zooming in* on a function for which the limit at a point exists.

Have the function $h(x) = \dfrac{3x^2 + 3x}{9x^2 + 11x + 2}$ entered in the Y1 location of the Y= list. Suppose that we want to estimate $\displaystyle\lim_{x \to -1} h(x)$.

Draw a graph of *h* with ZOOM 4 [ZDecimal]. Press ZOOM 2 [Zoom In] and use ◄ and ▲ to move the blinking cursor until you are near the point on the graph where $x = -1$. Press ENTER . If you look closely, you can see the hole in the graph at $x = -1$. (Note that we are *not* tracing the graph of *h*.)	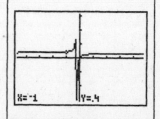

<table>
<tr><td>If the view is not magnified enough to see what is happening around $x = -1$, have the cursor near the point on the curve where $x = -1$ and press ENTER to zoom in again.</td><td>X=-1   Y=.35</td><td>X=-1   .Y=.3875</td></tr>
<tr><td>To estimate $\lim\limits_{x \to -1} h(x)$, press TRACE, use ▶ and ◀ to trace the graph close to, and on either side of $x = -1$, and observe the sequence of y-values.</td><td>Y1=(3X2+3X)/(9X2+11X+2)

X=-1.00625 .Y=.42781222</td><td>Y1=(3X2+3X)/(9X2+11X+2)

X=-.99375   .Y=.42934293</td></tr>
</table>

Observing the sequence of y-values is the same procedure as numerically estimating the limit at a point. Therefore, it is not the value at $x = -1$ that is important; the limit is what the output values displayed on the screen approach as x approaches -1. It appears that $\lim\limits_{x \to -1^-} h(x) \approx 0.43$

and $\lim\limits_{x \to -1^+} h(x) \approx 0.43$. Therefore, we conclude that $\lim\limits_{x \to -1} h(x) \approx 0.43$.

 ## 1.4 Linear Functions and Models

This portion of the *Guide* gives instructions for entering real-world data into the calculator and finding familiar function curves to fit that data. You will use the beginning material in this section throughout all the chapters in *Calculus Concepts*.

CAUTION: Be very careful when you enter data in your calculator because your model and all of your results depend on the values that you enter! Always check your entries.

1.4.1a ENTERING DATA We illustrate data entry using the values in Table 1.19 in Section 1.4 of *Calculus Concepts*:

Year	1999	2000	2001	2002	2003	2004
Tax (dollars)	2532	3073	3614	4155	4696	5237

Press STAT 1 [EDIT] to access the 6 lists that hold data. You see only the first 3 lists, (L1, L2, and L3) but you can access the other 3 lists (L4, L5, and L6) by having the cursor on the list name and pressing ▶ several times. If you do not see these list names, return to the statistical setup instructions on page A-1 of this *Guide*.

In this text, we usually use list L1 for the input data and list L2 for the output data. If there are any data values already in your lists, see Section 1.4.1c of this *Guide* and first delete any "old" data. To enter data in the lists, do the following:

<table>
<tr><td>Position the cursor in the first location in list L1. Enter the input data into list L1 by typing the years from top to bottom in the L1 column, pressing ENTER after each entry.

After typing the sixth input value, 2004, use ▶ to cause the cursor go to the top of list L2. Enter the output data in list L2 by typing the entries from top to bottom in the L2 column, pressing ENTER after each tax value.</td><td>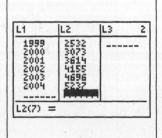</td></tr>
</table>

1.4.1b EDITING DATA   If you incorrectly type a data value, use the cursor keys (*i.e.,* the arrow keys ▶, ◀, ▲, and/or ▼) to darken the value you wish to correct and then type the correct value. Press ENTER.

- To *insert* a data value, put the cursor over the value that will be directly below the one you will insert, and press 2nd DEL (INS). The values in the list below the insertion point move down one location and a 0 is filled in at the insertion point. Type the data value to be inserted over the 0 and press ENTER. The 0 is replaced with the new value.

- To *delete* a single data value, move the cursor over the value you wish to delete and press DEL. The values in the list below the deleted value move up one location.

1.4.1c DELETING OLD DATA   Whenever you enter new data in your calculator, you should first delete any previously entered data. There are several ways to do this, and the most convenient method is illustrated below.

Access the data lists with STAT 1 [EDIT]. (You probably have different values in your lists if you are deleting "old" data.) Use ▲ to move the cursor over the name L1. Press CLEAR ENTER.	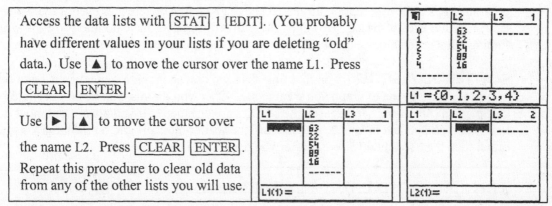
Use ▶ ▲ to move the cursor over the name L2. Press CLEAR ENTER. Repeat this procedure to clear old data from any of the other lists you will use.	

1.4.1d FINDING FIRST DIFFERENCES   When the input values are evenly spaced, you can use program DIFF to compute first differences in the output values. Program DIFF is given in the *TI-83 Program Appendix* at the *Calculus Concepts* web site. Consult the Programs category in *TroubleShooting the TI-83* in this *Guide* if you have questions about obtaining the programs.

Have the data given in Table 1.19 in Section 1.4 of *Calculus Concepts* entered in your calculator. (See Section 1.4.1a of this *Guide*.) Exit the list menu with 2nd MODE (QUIT).	L1 / L2 / L3 2 1999 2532 ------ 2000 3073 2001 3614 2002 4155 2003 4696 2004 5237 L2(7) =
To run the program, press PRGM followed by the number that is to the left of the DIFF program location, and press ENTER. The message on the right appears on your screen. (*Note:* We use the information in lists L4 and L5 in the next chapter.)	HAVE X IN L1 HAVE Y IN L2—SEE 1ST DIFF IN L3, 2ND DIFF IN L4, PERCENTAGE DIFFS IN L5 Done
Press STAT 1 [EDIT] to view the first differences in list L3. The first differences in L3 are constant at 541, so a linear function gives a perfect fit to these tax data.	L1 / L2 / L3 3 1999 2532 541 2000 3073 541 2001 3614 541 2002 4155 541 2003 4696 541 2004 5237 ------ ------ ------ L3 ={541,541,541...

You will find program DIFF very convenient to use in the next chapter because of the other options it has. However, if you do not want to use program DIFF at this time, you can use a built-in capability of your TI-83 to compute the first differences of any list. Be certain that you have the output data entered in list L2 in your calculator.

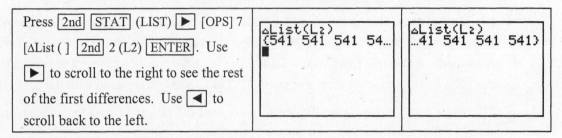

Press 2nd STAT (LIST) ▶ [OPS] 7 [ΔList ( ] 2nd 2 (L2) ENTER. Use ▶ to scroll to the right to see the rest of the first differences. Use ◀ to scroll back to the left.	ΔList(L₂) {541 541 541 54… ▮	ΔList(L₂) …41 541 541 541}

NOTE: Program DIFF **should not** be used for data with input values (entered in L1) that are *not* evenly spaced. First differences give no information about a possible linear fit to data with inputs that are not the same distance apart. If you try to use program DIFF with input data that are not evenly spaced, the message INPUT VALUES NOT EVENLY SPACED appears and the program stops. The ΔList option discussed above gives first differences even if the input values are not evenly spaced. However, these first differences have no interpretation when finding a linear function to fit data if the input values are not evenly spaced.

1.4.2a SCATTER PLOT SETUP   The first time that you draw a graph of data, you need to set the TI-83 to draw the type of graph you want to see. Once you do this, you never need to do this set up again (unless for some reason the settings are changed). If you always put input data in list L1 and output data in list L2, you can turn the scatter plots off and on from the Y= screen rather than the STAT PLOTS screen after you perform this initial setup.

Press 2nd Y= (STAT PLOT) to display the STAT PLOTS screen. (Your screen may not look exactly like this one.)	STAT PLOTS 1:Plot1…Off L₁ L₂ 2:Plot2…Off L₃ L₄ 3:Plot3…Off L₅ L₆ 4↓PlotsOff
On the STAT PLOTS screen, press 1 to display the Plot1 screen, press ENTER to turn Plot1 on, and then highlight each option shown on the right, pressing ENTER to choose it. (You can choose any of the 3 marks at the bottom of the screen.)	Plot1 Plot2 Plot3 On Off Type: Xlist:L₁ Ylist:L₂ Mark: ▫ + ·
Press Y= and notice that Plot1 at the top of the screen is darkened. This tells you that Plot1 is turned on and ready to graph data. Press 2nd MODE (QUIT) to return to the home screen.	Plot1 Plot2 Plot3 \Y1=▮ \Y2= \Y3= \Y4= \Y5= \Y6= \Y7=

- A scatter plot is *turned on* when its name on the Y= screen is darkened. To turn Plot1 off, use ▲ to move the cursor to the Plot1 location, and press ENTER. Reverse the process to turn Plot1 back on.

- TI-83 lists can be named and stored in the calculator's memory for later recall and use. Refer to Sections 1.4.3c and 1.4.3d of this *Guide* for instructions on storing data lists and later recalling them for use.

CAUTION: Any time that you enter the name of a numbered list (for instance: L1, L2, and so forth), you should use the TI-83 symbol for the name, **not** a name that you type with the alphabetic and numeric keys. For example, to enter the list name L1 on the TI-83, press 2nd 1 (L1). If you instead entered L1 with the keystrokes ALPHA)) (L) 1, you would create a new list called L1 that the calculator would not recognize as the list containing data that you entered while in the STAT [EDIT] mode.

The screen to the right contains three entries marked L1. The first is the statistical list L1 that contains input data. See what is in your L1 list by pressing 2nd 1 (L1) ENTER .

The second entry is obtained by pressing ALPHA)) (L) 1 ENTER . Whatever constant is stored in memory location L multiplied by 1 is what you see printed on the screen.

If you stored numbers in a list that you created and named L1 using ALPHA)) (L) 1 ENTER , you will see the third entry shown on the screen to the right when you press 2nd STAT (LIST) followed by the number corresponding to the location of L1 and ENTER .

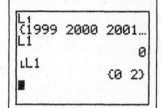

Your list L1 may not contain the same values as the one shown above. You may not have the third entry shown above in your calculator. If you do have it, it will contain different numbers.

If you have the third entry that is shown on the screen above (ʟL1) or if you see any lists that begin with the letter L followed by a large number (not a small number which denotes a built-in list) when you press 2nd STAT (LIST), refer to Section 1.4.3e of this *Guide*. These lists need to be deleted or you will probably be using incorrect data for everything!

1.4.2b DRAWING A SCATTER PLOT OF DATA   Any functions that are turned on in the Y= list will graph when you plot data. Therefore, you should clear or turn them off before you draw a scatter plot. We illustrate how to graph data using the modified tax data that follows Example 2 in Section 1.4 of *Calculus Concepts*.

Year	1999	2000	2001	2002	2003	2004
Tax (in dollars)	2541	3081	3615	4157	4703	5242

Press Y= . If any entered function is no longer needed, clear it by moving the cursor to its location and pressing CLEAR . If you want the function(s) to remain but not graph when you draw the scatter plot, refer to Section 1.2.2b of this *Guide* for instructions on how to turn function(s) off. Also be sure that Plot 1 on the top left of the Y= screen is darkened.

Press STAT 1 [EDIT]. Using the table given above, enter the year data in L1 and the modified tax data in L2 according to the instructions given in Sections 1.4.1a-c of this *Guide*.

(You can either leave values in the other lists or clear them.)

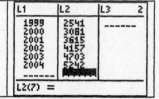

Press $\boxed{\text{ZOOM}}$ 9 [ZoomStat] to have the calculator set an auto-scaled view of the data and draw the scatter plot.

Note that ZoomStat does not reset the *x*- and *y*-axis tick marks. You should do this manually with the WINDOW settings if you want different spacing between the tick marks.)

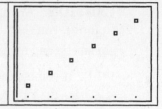

Recall that if the data are perfectly linear (that is, every data point falls on the graph of a line), the first differences in the output values are constant. The first differences for the original tax data were constant at $541, so a linear function fit the data perfectly. What information is given by the first differences for these modified tax data?

Run program DIFF by pressing $\boxed{\text{PRGM}}$ followed by the number that is to the left of the DIFF program location and press $\boxed{\text{ENTER}}$. Press $\boxed{\text{STAT}}$ 1 [EDIT] to view the first differences in list L3.

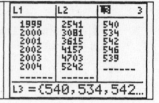

These first differences are close to being constant. This information, together with the linear pattern shown by the scatter plot, are a good indication that a linear function is likely to give a good fit to the data.

1.4.2c FINDING A LINEAR FUNCTION TO MODEL DATA

Throughout this course, you will often have your TI-83 find a function that best fits a set of data. Your calculator can find two different, but equivalent, forms of a linear function: $y = ax + b$ or $y = a + bx$. For convenience, we always choose the function with equation $y = ax + b$.

| Press $\boxed{\text{2nd}}$ $\boxed{\text{MODE}}$ (QUIT) to return to the home screen. Then, press $\boxed{\text{STAT}}$ $\boxed{\blacktriangleright}$ [CALC] 4 [LinReg(ax+b)] to access the TI-83's "fit a linear function to the data" instruction. | ```
EDIT CALC TESTS
1:1-Var Stats
2:2-Var Stats
3:Med-Med
4▮LinReg(ax+b)
5:QuadReg
6:CubicReg
7↓QuartReg
``` |
| Copy the fit instruction to the home screen by pressing $\boxed{4}$. At the same time, paste the equation into the Y1 location of the Y= list by pressing $\boxed{\text{VARS}}$ $\boxed{\blacktriangleright}$ [Y–VARS] 1 [Function] 1 [Y1]. Press $\boxed{\text{ENTER}}$. | `LinReg(ax+b) Y₁`   `LinReg`
`y=ax+b`
`a=540.3714286`
`b=-1077663.581` |

CAUTION: The best-fit function found by the calculator is also called a *regression function*. The coefficients of the regression function **never should be rounded** when you are going to use it! This is not a problem because the calculator pastes the entire equation it finds into the Y= list at the same time the function is found if you follow the instructions given above.

The linear equation of best fit for the modified tax data that was entered into lists L1 and L2 in Section 1.4.2b and displayed on the home screen has also been copied into Y1. Press $\boxed{\text{Y=}}$ to verify this.

```

NOTE: When finding the equation of best fit, the TI-83 will by default use L1 as the input data list and L2 as the output data list unless you tell it do otherwise. It is possible to use lists other than L1 and L2 for the input and output data. However, if you do so, you must set one of the STAT PLOT locations to draw the scatter plot for those other lists (as described in Section 1.4.2b). To find the best-fit function, replace L1 and L2 by the other lists in the fit instruction. To paste the function into a location other than Y1, just change Y1 in the instructions given above to the number corresponding to the location of the function you want. For example, these keystrokes find a linear function for input data in list L3 and output data in list L4 and paste the equation into graphing location Y2: |STAT| |▶| [CALC] 4 [LinReg(ax+b)]

|2nd| 3 (L3) |,| |2nd| 4 (L4) |,| |VARS| |▶| [Y–VARS] 1 [Function] 2 [Y2] |ENTER|.

CAUTION: If you see a number that is labeled *r* on the screen that shows the linear function you fit to the data, tell the TI-83 to not show this value again by pressing (on the home screen) |2nd| 0 (CATALOG) |x⁻¹| and press |▼| until you the cursor arrow is next to DiagnosticOff.

Press |ENTER| |ENTER|. The *r* that is shown is called the *correlation coefficient*. This and a quantity called r^2, the *coefficient of determination*, are numbers that you will learn about in a statistics course. It is not appropriate[3] to use these values in a calculus course.

Graphing the Line of Best Fit: After finding a best-fit equation, you should always draw the graph of the function on a scatter plot to verify that the function gives a good fit to the data.

Press |GRAPH| to overdraw the function you pasted in the Y= list on the scatter plot of the data. 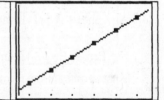
(As we suspected from looking at the scatter plot and the first differences, this function provides a very good fit to the data.)

1.4.2d COPYING A GRAPH TO PAPER   Your instructor may ask you to copy what is on your graphics screen to paper. If so, use the following ideas to more accurately perform this task.

After using a ruler to place and label a scale (*i.e.,* tick marks) on your paper, use the trace values (as shown below) to draw a scatter plot and graph of the line on your paper.

Press |GRAPH| to return the modified tax data graph found in Section 1.4.2c to the screen. Press |TRACE| and |▶|. The symbols P1:L1,L2 in the upper left-hand corner of the screen indicate that you are tracing the scatter plot of the data in Plot 1.	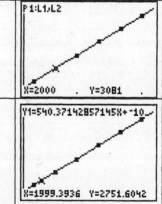
Press |▼| to move the trace cursor to the linear function graph. The equation at the top of the screen is that of the function that you are tracing (in this case, Y1). Use |▶| and/or |◀| to locate values that are as "nice" as possible and mark those points on your paper. Use a ruler to connect the points and draw the line.	

---

[3] Unfortunately, there is no single number that can be used to tell whether one function better fits data than another. The correlation coefficient only compares linear fits and should not be used to compare the fits of different types of functions. For the statistical reasoning behind this statement, read the references in footnote 6 on page A-30.

- If you are copying the graph of a continuous curve rather than a straight line, you need to trace as many points as necessary to see the shape of the curve while marking the points on your paper. Connect the points with a smooth curve.

1.4.3a ALIGNING DATA We return to the modified tax data entered in Section 1.4.2b. If you need L1 to contain the number of years after a certain year instead of the actual year, you need to *align* the input data. In this illustration, we shift all of the data points to 3 different positions to the left of where the original values are located.

Press STAT 1 [EDIT] to access the data lists. To copy the contents of one list to another list; for example, to copy the contents of L1 to L3, use ▲ and ▶ to move the cursor so that L3 is highlighted. Press 2nd 1 (L1) ENTER.	L1 L2 ▓ 3 1999 2541 2000 3081 2001 3615 ------ 2002 4157 2003 4703 2004 5242 ------ ------ L3 =L1▮

NOTE: This first step shown above is not necessary, but it will save you the time it takes to re-enter the input data if you make a mistake. Also, it is not necessary to first clear L3. However, if you want to do so, have the symbols L3 highlighted and press CLEAR ENTER.

To align the input data as the number of years past 1999, first press the arrow keys (◀ and ▲) so that L1 is highlighted. Tell the TI-83 to subtract 1999 from each number in L1 with 2nd 1 (L1) − 1999.	▓ L2 L3 1 1999 2541 1999 2000 3081 2000 2001 3615 2001 2002 4157 2002 2003 4703 2003 2004 5242 2004 ------ ------ ------ L1 =L1−1999▮
Press ENTER. Instead of an actual year, the input now represents the number of years since 1999. Return to the home screen with 2nd MODE (QUIT).	L1 L2 L3 1 0 2541 1999 1 3081 2000 2 3615 2001 3 4157 2002 4 4703 2003 5 5242 2004 ------ ------ ------ L1(1)=0
Find the linear function by pressing 2nd ENTER (ENTRY) as many times as needed until you see the linear regression instruction. To enter this function in a different location, say Y2, press ◀ and then press VARS ▶ [Y−VARS] 1 [Function] 2 [Y2].	LinReg(ax+b) Y2▮
Press ENTER and then press Y= to see the function pasted in the Y2 location. *Note:* If you want the aligned function to be in Y1, do not replace Y1 with Y2 before pressing ENTER to find the equation.	Plot1 Plot2 Plot3 \Y1▉540.37142857 145X+-1077663.58 09525 \Y2▉540.37142857 145X+2538.904761 9047 \Y3=
To graph this equation on a scatter plot of the aligned data, first turn off the function in Y1 (see page A-14 of this *Guide*) and then return to the home screen with 2nd MODE (QUIT). Press ZOOM 9 [ZoomStat].	

If you now want to find the linear function that best fits the modified tax data using the input data aligned another way, say as the number of years after 1900, first return to the data lists with STAT 1 [EDIT] and highlight L1.

Add 99 to each number currently in L1 with 2nd 1 (L1) + 99. Press ENTER. Instead of an actual year, the input now represents the number of years since 1900.	(screen showing L1, L2, L3 tables; bottom: L1 =L1+99 and L1(1)=99)

- There are many ways that you can enter the aligned input into L1. One method that you may prefer is to start over from the beginning. Replace L1 with the contents of L3 by highlighting L1 and pressing 2nd 3 (L3) ENTER. Once again highlight the name L1 and subtract 1900 from each number in L1 with 2nd 1 (L1) − 1900.

On the home screen, find the linear function for the aligned data by pressing 2nd ENTER (ENTRY) until you see the linear regression instruction. To enter this new equation in a different location, say Y3, press ◄ and then press VARS ► [Y–VARS] 1 [Function] 3 [Y3]. Press ENTER and then press Y= to see the function pasted in the Y3 location.	

- To graph this equation on a scatter plot of the aligned data, first turn off the other functions and then return to the home screen with 2nd MODE (QUIT). Press ZOOM 9 [ZoomStat].

1.4.3b USING A MODEL FOR PREDICTIONS
You can use one of the methods described in Section 1.1.1e or Section 1.1.1f of this *Guide* to evaluate the linear function at the indicated input value. Remember, if you have aligned the data, the input value at which you evaluate the function may not be the value given in the question you are asked.

CAUTION: Remember that you should always use the full model, *i.e.*, the function you pasted in the Y= list, not a rounded equation, for all computations.

Using the function in Y1 (the input is the year), in Y2 (the input is the number of years after 1999), or in Y3 (the input is the number of years after 1900), we predict that the tax owed in 2006 is approximately $6322.	Y1(2006)
	        6321.504762
	Y2(7)
	        6321.504762
	Y3(106)
	        6321.504762

You can also predict the tax in 2006 using the TI-83 table (with ASK chosen in TBLSET) and any of the 3 models found in the previous section of this *Guide*. As seen to the right, the predicted tax is approximately $6322.	(table with X, Y1 columns; 2006; Y1=6321.5047618)

1.4.3c NAMING AND STORING DATA
You can name data (either input, output, or both) and store it in the calculator memory for later recall. You may or may not want to use this feature. It will be helpful if you plan to use a large data set several times and do not want to reenter the data each time.

To illustrate the procedure, let's name and store the modified tax output data that was entered in Section 1.4.2b.

Press [2nd] [MODE] (QUIT) to return to the home screen. You can view any list from the STAT EDIT mode (where the data is originally entered) or from the home screen by typing the name of the list and pressing [ENTER]. View the modified tax data in L2 with [2nd] 2 (L2) [ENTER].	L₂ {2541 3081 3615…
Pressing [▶] allows you to scroll through the list to see the portion that is not displayed. To store this data with the name TAX, press [2nd] 2 (L2) [STO▶] [2nd] [ALPHA] T A X [ENTER].	L₂ {2541 3081 3615… L₂→TAX {2541 3081 3615… ■

CAUTION: Do not store data to a name that is routinely used by the TI-83. Such names are L1, L2, ..., L6, MATH, LOG, MODE, A, B, and so forth. Note that if you use a single letter as a name, this might cause one or more of the programs to not execute properly.

1.4.3d **RECALLING STORED DATA**   The data you have stored remains in the memory of the TI-83 until you delete it using the instructions given in Section 1.4.3e of this *Guide*. When you wish to use the stored data, recall it to one of the lists L1, L2, ..., L6. We illustrate with the list named TAX, which we stored in L2. Press [2nd] [STAT] (LIST) and under NAMES, find TAX.

Press the number* corresponding to the location of the list. Press [STO▶] [2nd] 2 (L2) [ENTER]. List L2 now contains the TAX data.	**NAMES** OPS MATH 1:RESID **2:**TAX	LTAX→L₂ {2541 3081 3615…

*An equivalent action is to use [▼] to highlight the number (here, 2) and press [ENTER].

1.4.3e **DELETING USER-STORED DATA**   You do not need to delete any data lists unless your memory is getting low or you just want to do it. One exception to this is that you should delete any list that begins with the letter *L* and ends in a single *large* (*i.e.,* a large-size type) number. If you see a list beginning with *L* (such as L1 on the first screen below), you should delete it because the TI-83 may use the data in this list instead of the data you enter in the first list in the STAT EDIT mode. For illustration purposes, we delete lists L1 and TAX.

Press [2nd] [STAT] (LIST) to view the user-defined lists. Then press [2nd] [+] (MEM) 2 [Delete].	**NAMES** OPS MATH **1:**L1 2:RESID 3:TAX	**MEMORY** 1:Check RAM… **2:**Delete… 3:Clear Entries 4:ClrAllLists 5:Reset…
Next, press 4: [List] and use [▼] to move the cursor opposite L1 (or TAX). DO NOT delete the TI-83's list L1 (*L* followed by a small, not large, 1). If you did not see a list L1 when [2nd] [STAT] (LIST) was pressed, you only have the list TAX to delete.		DELETE:List L₃ 63 L₄ 45 L₅ 54 L₆ 54 ▶L1 28 RESID 67 TAX 65

Press ENTER. To delete another list, use ▼ or ▲ to move the cursor opposite that list name and press ENTER. Exit this screen with 2nd MODE (QUIT) when finished.

```
DELETE:List
 L₃           63
 L₄           45
 L₅           54
 L₆           54
▶RESID        67
```

WARNING: Be careful when in the DELETE menu. Once you delete something, it is gone from the TI-83 memory and cannot be recovered. If you mistakenly delete one of the lists L1, L2, …, L6, you need to redo the statistical setup as indicated on page A-1 of this *Guide*.

1.4.4 **WHAT IS "BEST FIT"?**

It is important to understand the method of least squares and the conditions necessary for its use if you intend to find the equation that best-fits a set of data. You can explore the process of finding the line of best fit with program LSLINE. (Program LSLINE is given in the *TI-83 Program Appendix*.) For your investigations of the least-squares process with this program, it is better to use data that is not perfectly linear and data for which you do *not* know the best-fitting line.

We use the data in the table below to illustrate program LSLINE, but you may find it more interesting to input some other data[4].

x (input in L1)	1	3	7	9	12
y (output in L2)	1	6	10	16	20

Before using program LSLINE, clear all functions from the Y= list, turn on Plot1 by darkening the name Plot1 on the Y= screen, and enter your data in lists L1 and L2. (If Plot 1 is not turned on, the program will not execute properly.) Next, draw a scatter plot with ZOOM 9 [ZoomStat]. Press WINDOW and reset Xscl and Yscl so that you can use the tick marks to help identify points when you are asked to give the equation of a line to fit the data. (Four the *Guide* illustration, we set Xscl to 1 and Yscl to 5.) Press GRAPH, view the scatter plot, and then return to the home screen.

To activate program LSLINE, press PRGM followed by the number corresponding to the location of the program, and press ENTER. The program first displays the scatter plot you constructed and pauses for you to view the screen.

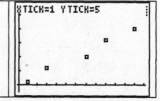

NOTE: While the program is calculating, there is a small vertical line in the upper-right hand corner of the graphics screen that is dashed and "wiggly". This program pauses several times during execution for you to view the screen. Whenever the program pauses, the small line is "still" and you should press ENTER to resume execution after you have looked at the screen.

The program next asks you to find the *y*-intercept and slope of *some* line you estimate will go "through" the data. (You should not expect to guess the best-fit line on your first try!)

---

[4] This program works well with approximately 5 data points. Interesting data to use in this illustration are the height and weight, the arm span length and the distance from the floor to the navel, or the age of the oldest child and the number of years the children's parents have been married for 5 randomly selected persons.

Use the tick marks to estimate rise divided by run and note a possible *y*-intercept. After pressing ENTER to resume the program, enter your guess for the slope and *y*-intercept.

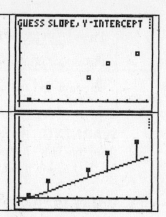

After entering your guess for the *y*-intercept, your line is drawn and the errors are shown as vertical line segments on the graph. (You may have to wait a moment to see the vertical line segments.) Press ENTER to continue the program.

Next, the sum of squared errors, SSE, is displayed for your line. Decide whether you want to move the *y*-intercept of the line or change its slope to improve the fit to the data.

Press ENTER and enter 1 to choose the TRY AGAIN? option. After again viewing the errors, enter another guess for the *y*-intercept and/or slope. The process of viewing your line, the errors, and display of SSE is repeated.

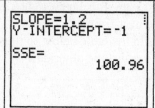

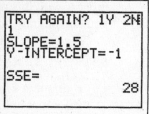

If the new value displayed for SSE is smaller than the value of SSE for your first guess, you have improved the fit.

When you feel an SSE value close to the minimum value is found, enter 2 at the TRY AGAIN? prompt. The program then overdraws the line of best fit on the graph and shows the errors for the line of best fit.

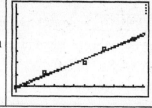

The program ends by displaying the coefficients *a* and *b* of the best-fit line $y = ax + b$ as well as the minimum SSE. Press 2nd MODE (QUIT) to end the program. Use program LSLINE to explore[5] the method of least squares[6] that the TI-83 uses to find the line of best fit.

---

[5] Program LSLINE is for illustration purposes only. Actually finding the line of best fit for a set of data should be done according to the instructions in Section 1.4.2c of this *Guide*.

[6] Two articles that further explain "best-fit" are H. Skala, "Will the Real Best Fit Curve Please Stand Up?" Classroom Computer Capsule, *The College Mathematics Journal*, vol. 27, no. 3, May 1996 and Bradley Efron, "Computer-Intensive Methods in Statistical Regression," *SIAM Review*, vol. 30, no. 3, September 1988.

Chapter 2   Ingredients of Change: Nonlinear Models

 2.1 Exponential Functions and Models

As we begin to consider functions that are not linear, it is very important that you be able to draw scatter plots, find numerical changes in output data, and recognize the underlying shape of the basic functions to be able to identify which function best models a particular set of data. Finding the model is only a means to an end – being able to use mathematics to describe the changes that occur in real-world situations.

2.1.1a ENTERING EVENLY SPACED INPUT VALUES   When an input list consists of many values that are the same distance apart, there is a calculator command that will generate the list so that you do not have to type the values in one by one. The syntax for this sequence command is *seq(formula, variable, first value, last value, increment)*. When entering years that differ by 1, the formula is the same as the variable and the increment is 1. Any letter can be used for the variable -- we choose to use X.   We illustrate the sequence command use to enter the input for the population data in Table 2.2 in Section 2.1 of *Calculus Concepts*.

Year	1994	1995	1996	1997	1998	1999	2000	2001	2002	2003
Population	7290	6707	6170	5677	5223	4805	4420	4067	3741	3442

Clear any old data from lists L1 and L2. To enter the input data, position the cursor in the first list so that L1 is darkened.

Generate the list of years beginning with 1994, ending with 2003, and differing by 1 with [2nd] [STAT] (LIST) [▶] [OPS] 5 [seq(] [X,T,θ,n] [,] [X,T,θ,n] [,] 1994 [,] 2003 [,] 1 [)].

Press [ENTER] and the sequence of years is pasted into L1. Manually enter the output values in L2.

2.1.1b FINDING PERCENTAGE DIFFERENCES   When the input values are evenly spaced, use program DIFF to compute the percentage differences in the output data. If the data are perfectly exponential (that is, every data point falls on the exponential function), the percentage change (which in this case would equal the percentage differences) in the output is constant. If the percentage differences are close to constant, this is an indication that an exponential model *may* be appropriate.

Return to the home screen. Find the percentage differences in the population data with program DIFF. Press [PRGM] and then the number that is to the left of the DIFF program location.

Press [ENTER] and the message on the right appears on your screen. Press [STAT] 1 [EDIT] and [▶] [▶] [▶] to view the percentage differences in list L5.

Scroll down L5 with ▼ to see all 9 values. The percentage differences are close to constant, so an exponential function will probably give a good fit to these data.

2.1.2 FINDING AN EXPONENTIAL FUNCTION TO MODEL DATA

Use your calculator to find an exponential equation to model the small town population data. The exponential function we use is of the form $y = ab^x$. It is very important that you align large numbers (such as years) whenever you find an exponential model. If you don't align to smaller values, the TI-83 may return an error message or an incorrect function due to computation errors.

We continue to use the data in Table 2.2 of the text. First, align the input data to smaller values. Other alignments are possible, but we choose to align the data so that $x = 0$ in 1994. To align, darken the list name and press 2nd 1 (L1) − 1994 ENTER.	**L1** **L2** **L3** **1** 1994 7290 ⁻583 1995 6707 ⁻537 1996 6170 ⁻493 1997 5677 ⁻454 1998 5223 ⁻418 1999 4805 ⁻385 2000 4420 ⁻353 L1 =L1−1994
Return to the home screen. Following the same procedure that you did to find a linear function, find the exponential function <u>and</u> paste the equation into the Y1 location of the Y= list by pressing STAT ▶ [CALC] 0 [ExpReg] VARS ▶ [Y–VARS] 1 [Function] 1 [Y1].	ExpReg Y1■
Press ENTER to find the equation and paste it into the Y1 location. Press Y= to view Y1. Press ZOOM 9 [ZoomStat] to draw the scatter plot and the graph of the function.	**Plot1** Plot2 Plot3 \Y1■7290.2503151 852*.9199949014 276^X \Y2= \Y3= \Y4= \Y5=
As the percentage differences indicate, the function gives a very good fit for the data. To estimate the town's population in 2004, remember that x is the number of years past 1994 and evaluate the function in Y1 at $x = 10$. We predict the population to be about 3167 people in 2004.	a=7290.250315 b=.9199949014 Y1(10) 3166.625068

- Do not confuse the percentage differences found from the data with the percentage change for the exponential function. The constant percentage change in the exponential function $y = ab^x$ is $(b - 1)100\%$. The constant percentage change for the function is a single value whereas the percentage differences calculated from the data are many different numbers.

2.2 Logarithmic Functions and Models

In this section we consider another function that can be used to fit data – the log function. This function is the inverse function for the exponential function discussed in the previous section. We recognize when to use this function by considering the behavior of the data rather than a numerical test involving differences.

2.2.1 FINDING A LOG FUNCTION TO MODEL DATA

Use your calculator to find a log equation of the form $y = a + b \ln x$. We illustrate finding this function with the air pressure and altitude data in Table 2.5 in Section 2.2 of *Calculus Concepts*.

Air pressure (inches of mercury)	13.76	5.56	2.14	0.82	0.33
Altitude (thousands of feet)	20	40	60	80	100

Clear the data that is currently in lists L1 and L2. Enter the air pressure data in L1 and the altitude data in L2. (If you wish, clear list L3, but it is not necessary to do so.)

Delete (with CLEAR) any functions that are in the Y= list. A scatter plot of the data drawn with ZOOM 9 [ZoomStat] shows the slow decline that can indicate a log model.	
As when modeling linear and exponential functions, find and paste the log equation into the Y1 location of the Y= list by pressing STAT ▶ [CALC] 9 [LnReg] VARS ▶ [Y–VARS] 1 [Function] 1 [Y1]. Press Y= to view the pasted function.	
Either press ZOOM 9 [ZoomStat] or GRAPH to view the function on the plot of the data. It appears to be an excellent fit.	

2.2.2 AN INVERSE RELATIONSHIP Exponential and log functions are inverse functions. If you find the inverse function for the log equation that models altitude as a function of air pressure (given in the section above), it will be an exponential function. We briefly explore this concept using the TI-83. Leave the log function found in the previous section ($y = 76.174 - 21.331 \ln x$) in the Y1 location of the Y= list because we use it again in this investigation.

Find the exponential equation that models air pressure as a function of altitude and paste it into the Y2 location with STAT ▶ [CALC] 0 [ExpReg] 2nd 2 (L2) , 2nd 1 (L1) , VARS ▶ [Y–VARS] 1 [Function] 2 [Y2]. Note that L2 is typed first!	
Press ENTER. Then press Y= and turn Plot 1 off. Enter the composite function Y1∘Y2 (or Y2∘Y1) in the Y3 location. Enter some values in the TABLE and see that what is returned for Y3 is close to the value of X that was entered.	We use the Composition Property of Inverse Functions.	Press ▶ to see the Y3 column.

- There is a slight difference in the exponential function and the true inverse function for Y1 because of rounding and because the exponential function in Y2 was not found algebraically but from fitting an equation to data points.

You can also graphically verify that Y1 and Y2 are inverse functions. Clear or turn off the composite function in Y3 and turn off Y2.

On the home screen, have the TI-83 draw the graph of the inverse of Y1 with [2nd] [PRGM] (DRAW) 8 (DrawInv) [VARS] [▶] [Y–VARS] 1 [Function] 1 [Y1] [ENTER] . Press [Y=] , turn Y2 on, and press [GRAPH] . You see the same graph, indicating that Y2 and the inverse function either are the same function or close enough to not be visually distinguished.

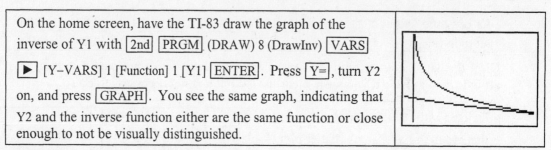

CAUTION: From this point on in the *Guide*, the keystrokes used to access function locations in the Y= list will not be printed. You should remember that you must type the function name using [VARS] [▶] [Y–VARS] 1 [Function] followed by the number of the desired function location. The TI-83 does not recognize [ALPHA] 1 (Y) as the name of a function in the Y= list.

2.3 Logistic Functions and Models

This section introduces the logistic function that can be used to describe growth that begins as exponential and then slows down to approach a limiting value. Logistic functions are often in applications to the spread of a disease or virus. The TI-83's random number generator can be used to collect data to illustrate such a situation.

2.3.1 GENERATING RANDOM NUMBERS Imagine all the real numbers between 0 and 1, including 0 but not 1, written on identical slips of paper and placed in a hat. Close your eyes and draw one slip of paper from the hat. You have just chosen a number "at random."

Your calculator doesn't offer you a random choice of all real numbers between 0 and 1, but it allows you to choose, *with an equal chance of obtaining each one*, any of 10^{14} different numbers between 0 and 1 with its random number generator called rand.

Before using your calculator's random number generator for the first time, you should "seed" the random number generator. (This is like mixing up all the slips of paper in the hat.)

Pick some number, <u>not</u> the one shown on the right, and store it as the "seed" with *your number* [STO▸] [MATH] [◀] [PRB] 1 [rand] [ENTER] . Everyone needs to have a different seed, or the number choices will not be random.	`2658→rand` `            2658`

Enter rand again, and press [ENTER] several times.
 	
Your list of random numbers should be different from the one on the right if you entered a different seed. Notice that all numbers are between 0 and 1.
	`rand`
	`     .7005331056`
	`     .9836308181`
	`     .0068461306`
	`     .2110053651`
	`     .2225253266`
	`      .728178859`

We usually count with the numbers 1, 2, 3, … and so forth, not with decimal values. To adapt the random number generator to choose a whole number between, and including, 1 and *N*, we need to enter the instruction int(N rand + 1).

Enter, using a specific number for *N*, [MATH] [▶] 5 [int(] *N* [MATH] [◀] [PRB] 1 [rand] [+] 1 [)] . For instance, to choose a computer at random from a room that contains 10 computers, use *N* = 10. Keep pressing [ENTER] to choose more numbers.	`int(10rand+1)`
	`            7`
	`            8`
	`            4`
	`            6`
	`           10`
	`           10`

NOTE: If you want to generate a list of five random numbers, change the last 1 in the above instruction to 5; if you want 10 random numbers, change it to 10, and so forth.

As, previously stated, your list of numbers should be different from those shown above. All values will be between 1 and 10. Note that it is possible to obtain the same value more than once. This corresponds to each slip of paper being put back in the hat after being chosen and the numbers mixed well before the next number is drawn.

2.3.2a FINDING A LOGISTIC FUNCTION TO MODEL DATA

You can use your calculator to find a logistic function of the form $y = \dfrac{L}{1 + Ae^{-Bx}}$. Unlike the other functions we have fit to data, the logistic equation that you obtain may be slightly different from a logistic equation that is found with another calculator. (Logistic equations in *Calculus Concepts* were found using a TI-83.) As you did when finding an exponential equation for data, large input values must be aligned or an error or possibly an incorrect answer could be the result.

Note that the TI-83 finds a "best-fit" logistic function rather than a logistic function with a limiting value L such that no data value is ever greater than L. We illustrate finding a logistic function using the data in Table 2.8 of Section 2.3 in *Calculus Concepts*.

Time (hours after 8 A.M.)	0	1	2	3	4	5
Total number of infected computers	1	2	4	6	8	9

Clear any old data. Delete any functions in the Y= list and turn on Plot 1. Enter the data in the above table with the hours after 8 A.M. in list L1 and the number of computers in list L2:

Construct a scatter plot of the data. An inflection point is indicated, and the data appear to approach limiting values on the left and on the right. A logistic model seems appropriate.	
To find the logistic function and paste the equation into the Y1 location of the Y= list, press [STAT] [▶] [CALC] [ALPHA] B [Logistic] [VARS] [▶] [Y–VARS] 1 [Function] 1 [Y1].	`Logistic Y1▮`

Note: TI-83 menu items can be chosen by using [▼] to scroll to the location of the item and pressing [ENTER]. For instance, instead of choosing Logistic by pressing [ALPHA] B as indicated in the instruction above, you can scroll down the list of functions until B is highlighted and press [ENTER].

Press [ENTER]. This equation will take longer to generate than the other equations.
 	
Note that the TI-83 uses the variable c for the value that the text calls L.
	`Logistic`
	`y=c/(1+ae^(-bx))`
	`a=9.338121425`
	`b=.8951126464`
	`c=9.986321858`

Press [GRAPH] (or [ZOOM] 9 [ZoomStat]) to see the graph of the function drawn on the scatter plot.	

The limiting value, in this context, is about 10 computers. The two horizontal asymptotes are $y = 0$ (the line lying along the x-axis) and the line $y \approx 10$. To see the upper asymptote drawn on a graph of the data, enter 10 in Y2 and press $\boxed{\text{GRAPH}}$.

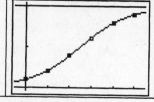

- Provided the input values are evenly spaced, program DIFF might be helpful when you are trying to determine if a logistic model is appropriate for certain data. If the first differences (in list L3 after running program DIFF) begin small, peak in the middle, and end small, this is an indication that a logistic model may provide a good fit to the data.

2.3.2b RECALLING MODEL PARAMETERS Rounding function parameters can often lead to incorrect or misleading results. You may find that you need to use the complete values of the coefficients after you have found a function that best fits a set of data. It would be tedious to copy all these digits in a long decimal number into another location of your calculator. You don't have to because you can recall any parameter found after you use one of the regressions.

We illustrate these ideas by recalling the parameter c for the logistic function found in the previous section of this *Guide*.

To recall the value of c in $y = c/(1 + ae^{\wedge}(-bx))$, press $\boxed{\text{VARS}}$ 5 [Statistics] $\boxed{\blacktriangleright}$ $\boxed{\blacktriangleright}$ [EQ] 4 (c).

```
XY Σ EQ TEST PTS
1:RegEQ
2:a
3:b
4:c
5:d
6:e
7↓r
```

Press $\boxed{\text{ENTER}}$ and the full value of c will be "pasted" where you had the cursor before pressing the keys that call up the value of c. (If you did not obtain the correct value for c, repeat finding the logistic function in the previous section and then repeat the previous instructions to find c.)

```
c
        9.986321858
```

- This same procedure applies to any equation you have found using the STAT CALC menu. For instance, if you had just fit an exponential function to data, the values of a and b in the EQ menu would be for that function and the other parameters would not apply.

WARNING: Once a different function is found, no parameters from a previous function fit to data remain stored in the calculator's memory.

2.3.3 VERTICALLY SHIFTING DATA If a scatter plot of data and/or a particular context indicate that an exponential or a logistic function is appropriate but the function does not appear to fit the data, it may be that a vertical shift of the data should be considered. We illustrate a vertical shift to improve the fit using the investment club data in Table 2.13 in Example 1 of Section 2.3 of *Calculus Concepts*:

Year	1990	1991	1992	1993	1994	1995	1996	1997	1998
Number of clubs	7085	7360	8267	10,033	12,429	16,054	25,409	31,828	36,500

First, clear your lists, and then enter the data in the table. Next, align the input data so that x represents the number of years since 1990. Draw a scatter plot of the data. The context of the situation and a scatter plot of the data indicate that a logistic model may be appropriate. (Note that it is necessary to align the input data to smaller values because of the exponential term in the logistic function.)

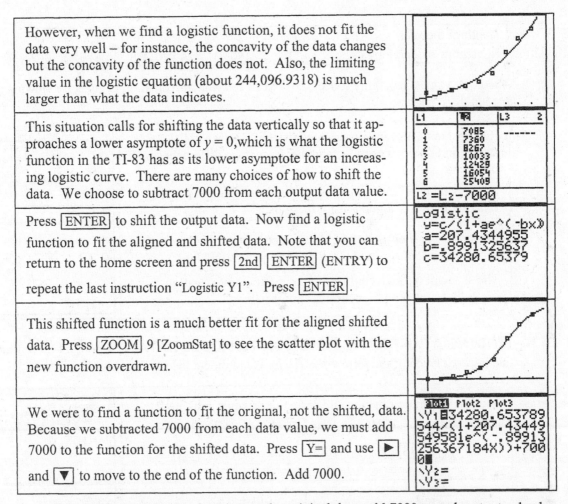

However, when we find a logistic function, it does not fit the data very well – for instance, the concavity of the data changes but the concavity of the function does not. Also, the limiting value in the logistic equation (about 244,096.9318) is much larger than what the data indicates.	
This situation calls for shifting the data vertically so that it approaches a lower asymptote of $y = 0$, which is what the logistic function in the TI-83 has as its lower asymptote for an increasing logistic curve. There are many choices of how to shift the data. We choose to subtract 7000 from each output data value.	L1 / L2 / L3 ... 0 7085 / 1 7360 / 2 8267 / 3 10033 / 4 12428 / 5 16054 / 6 25409 ... L2 =L2-7000
Press ENTER to shift the output data. Now find a logistic function to fit the aligned and shifted data. Note that you can return to the home screen and press 2nd ENTER (ENTRY) to repeat the last instruction "Logistic Y1". Press ENTER.	Logistic y=c/(1+ae^(-bx)) a=207.4344955 b=.8991325637 c=34280.65379
This shifted function is a much better fit for the aligned shifted data. Press ZOOM 9 [ZoomStat] to see the scatter plot with the new function overdrawn.	
We were to find a function to fit the original, not the shifted, data. Because we subtracted 7000 from each data value, we must add 7000 to the function for the shifted data. Press Y= and use ▶ and ▼ to move to the end of the function. Add 7000.	Plot1 Plot2 Plot3 \Y1=34280.653789 544/(1+207.43449 549581e^(-.89913 256367184X))+700 0 \Y2= \Y3=

- If you wish to graph this function on the original data, add 7000 to each output value by having the list name L2 darkened and pressing 2nd 2 (L2) + 7000 ENTER. Draw the graph with ZOOM 9 [ZoomStat].

NOTE: If you fit more than one shifted function and are unsure which of them better fits the data, zoom in on each near the "ends" of each graph and at the inflection points to see which function seems to best follow the pattern of the data.

2.4 Polynomial Functions and Models

You will in this section learn how to fit functions that have the familiar shape of a parabola or a cubic to data. Using your calculator to find these equations involves the same procedure as when using it to fit linear, exponential, log, or logistic functions.

2.4.1a FINDING SECOND DIFFERENCES When the input values are evenly spaced, you can use program DIFF to quickly compute second differences in the output values. If the data are perfectly quadratic (that is, every data point falls on a quadratic function), the second differences in the output values are constant. When the second differences are close to constant, a quadratic function *may* be appropriate for the data.

We illustrate these ideas with the roofing jobs data given on the first page of Section 2.4 in *Calculus Concepts*. We align the input data so that 1 = January, 2 = February, etc. Clear any old data and enter these data in lists L1 and L2:

Month of the year	1	2	3	4	5	6
Number of roofing jobs	117	140	224	368	575	842

The input values are evenly spaced, so we can see what information is given by viewing the second differences.

Run program DIFF and observe the first differences in list L3, the second differences in L4, and the percentage differences in list L5. The second differences are close to constant, so a quadratic function may give a good fit for these data.	L3: 23, 84, 144, 207, 267 — L4: 61, 60, 63, 60 — L5: 19.658, 60, 64.286, 56.25, 46.435 — L4(5) =
Construct a scatter plot of the data. (Don't forget to clear the Y= list of previously-used equations and to turn on Plot 1.) The shape of the data confirms the numerical investigation result that a quadratic function is appropriate.	(scatter plot)

2.4.1b FINDING A QUADRATIC FUNCTION TO MODEL DATA Use your TI-83 to find a quadratic function of the form $y = ax^2 + bx + c$ to model the roofing jobs data.

Find the quadratic function and paste the equation into the Y1 location of the Y= list by pressing [STAT] [▶] [CALC] 5 [QuadReg] [VARS] [▶] [Y–VARS] 1 [Function] 1 [Y1] [ENTER].	QuadReg y=ax²+bx+c a=30.57142857 b=-69.02857143 c=155.6
Press [GRAPH] to overdraw the graph of the function on the scatter plot. This function gives a very good fit to the data.	(graph)
How many jobs do we predict the company will have in August? Because August corresponds to $x = 8$, evaluate Y1 at $x = 8$. We predict that there will be 1559 jobs in August.	Y1(8) 1559.942857

2.4.2 FINDING A CUBIC FUNCTION TO MODEL DATA Whenever a scatter plot of data shows a single change in concavity, we are limited to fitting either a cubic or logistic function. If one or two limiting values are apparent, use the logistic equation. Otherwise, a cubic function should be considered. When appropriate, use your calculator to obtain the cubic function of the form $y = ax^3 + bx^2 + cx + d$ that best fits a set of data.

We illustrate finding a cubic function with the data in Table 2.18 in Example 3 of Section 2.4 in *Calculus Concepts*. The data give the average price in dollars per 1000 cubic feet of natural gas for residential use in the U.S. for selected years between 1980 and 2000.

Year	1980	1982	1985	1990	1995	1998	2000
Price (dollars)	3.68	5.17	6.12	5.80	6.06	6.82	7.71

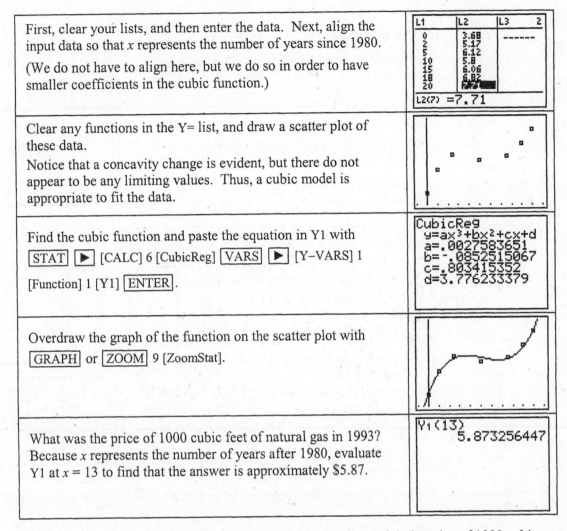

First, clear your lists, and then enter the data. Next, align the input data so that x represents the number of years since 1980. (We do not have to align here, but we do so in order to have smaller coefficients in the cubic function.)

Clear any functions in the Y= list, and draw a scatter plot of these data.

Notice that a concavity change is evident, but there do not appear to be any limiting values. Thus, a cubic model is appropriate to fit the data.

Find the cubic function and paste the equation in Y1 with STAT ▶ [CALC] 6 [CubicReg] VARS ▶ [Y–VARS] 1 [Function] 1 [Y1] ENTER.

Overdraw the graph of the function on the scatter plot with GRAPH or ZOOM 9 [ZoomStat].

What was the price of 1000 cubic feet of natural gas in 1993? Because x represents the number of years after 1980, evaluate Y1 at $x = 13$ to find that the answer is approximately $5.87.

Part b of Example 3 asks you to find when, according to the model, the price of 1000 cubic feet of natural gas first exceed $6. You can find the answer by using the SOLVER to find the solution to Y1 – 6 = 0. However, because we already have a graph of this function, the following may be easier. Note that if you do use the SOLVER, you should first TRACE the graph to estimate a guess for x that you need to enter in the SOLVER.

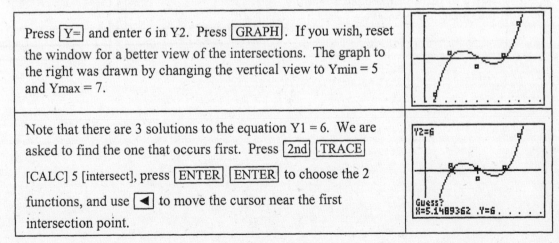

Press Y= and enter 6 in Y2. Press GRAPH. If you wish, reset the window for a better view of the intersections. The graph to the right was drawn by changing the vertical view to Ymin = 5 and Ymax = 7.

Note that there are 3 solutions to the equation Y1 = 6. We are asked to find the one that occurs first. Press 2nd TRACE [CALC] 5 [intersect], press ENTER ENTER to choose the 2 functions, and use ◀ to move the cursor near the first intersection point.

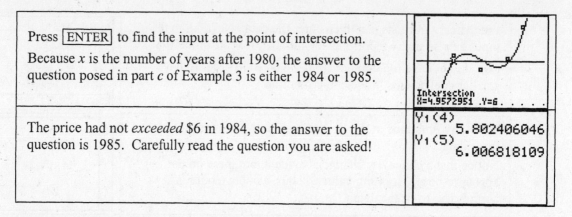

Press ENTER to find the input at the point of intersection. Because x is the number of years after 1980, the answer to the question posed in part c of Example 3 is either 1984 or 1985.

The price had not *exceeded* $6 in 1984, so the answer to the question is 1985. Carefully read the question you are asked!

Notice that the natural gas prices could have also been modeled with a cubic equation by renumbering the input data so that x is the number of years after 1900.

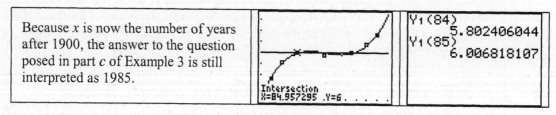

Return to the data and add 80 to each input value so that L1 represents the number of years after 1900 rather than the number of years after 1980.

Return to the home screen. Press 2nd ENTER (ENTRY) to return the instruction "Cubic Y1" to the screen. Press ENTER . Press Y= and have Y2 = 6.

Because the data are realigned, draw a graph of the equation with ZOOM 9 [ZoomStat] rather than GRAPH . Graphically finding where Y1 first exceeds 6 is done exactly as was previously discussed, but you need to carefully interpret the solution.

Because x is now the number of years after 1900, the answer to the question posed in part c of Example 3 is still interpreted as 1985.

Chapter 3   Describing Change: Rates

 3.1 Change, Percent Change, and Average Rates of Change

As you calculate average and other rates of change, remember that every numerical answer in a context should be accompanied by units telling how the quantity is measured. You should also be able to interpret each numerical answer. It is only through their interpretations that the results of your calculations will be useful in real-world situations.

3.1.1a FINDING AVERAGE RATE OF CHANGE   Finding an average rate of change using a function is just a matter of evaluating the equation at two different values of the input variable and dividing by the difference of those input values.

   We illustrate this concept using the function describing the population density of Nevada from 1960 through 2000 that is given in Example 3 of Section 3.1 of *Calculus Concepts*.

The population density referred to above is given by the function $p(t) = 0.1536(1.04892^t)$ people per square mile where t is the number of years after 1900. Press $\boxed{Y=}$, clear any functions, turn Plot 1 off, and enter this function in Y1 using X as the input variable.	`Plot1 Plot2 Plot3`
	`\Y1◼.1536(1.0489`
	`2^X)`
	`\Y2=`
	`\Y3=`
	`\Y4=`
	`\Y5=`
	`\Y6=`

To find the average rate of change of the population density from 1960 through 1980, first realize that 1960 corresponds to $x = 60$ and 1980 corresponds to $x = 80$. Then, recall that the average rate of change of p between 1960 and 1980 is given by the quotient $\dfrac{p(80)-p(60)}{80-60}$.

Return to the home screen and type the quotient $\dfrac{Y1(80)-Y1(60)}{80-60}$. Remember to enclose both the numerator and denominator of the fraction in parentheses.
 	
Finding this quotient in a single step avoids having to round intermediate calculation results.
	`(Y1(80)-Y1(60))/`
	`(80-60)`
	`        .2156842213`

Recall that rate of change units are output units per input units. On average, Nevada's population density increased about 0.22 person per square mile per year between 1960 and 1980.

To find the average rate of change between 1980 and 2000, recall the last expression with $\boxed{2nd}$ $\boxed{ENTER}$ (ENTRY) and replace 60 with 100 in two places. Press $\boxed{ENTER}$.	`(Y1(80)-Y1(60))/`
	`(80-60)`
	`        .2156842213`
	`(Y1(80)-Y1(100))`
	`/(80-100)`
	`        .5606162739`

NOTE: If you have many average rates of change to calculate, you could put the average rate of change formula in the graphing list: Y2 = (Y1(B) – Y1(A))/(B – A). Of course, you need to have a function in Y1. Then, on the home screen, store the inputs of the two points in A and B with 80 $\boxed{STO▸}$ $\boxed{ALPHA}$ A $\boxed{ALPHA}$ $\boxed{.}$ (:) 100 $\boxed{STO▸}$ $\boxed{ALPHA}$ B $\boxed{ENTER}$. All you need do next is type Y2 and press $\boxed{ENTER}$. Then, store the next set of inputs into A and/or B and recall Y2, using $\boxed{2nd}$ $\boxed{ENTER}$ (ENTRY) to recall each instruction. Press $\boxed{ENTER}$ and you have the average rate of change between the two new points. Try it!

3.1.1b CALCULATING PERCENTAGE CHANGE   You can find percentage changes using data either by the formula or by using program DIFF. To find a percentage change from a function instead of data, you should use the percentage change formula. We again use the population density of Nevada function in Example 3 of Section 3.1 of *Calculus Concepts* to illustrate.

Have $p(x) = 0.1536(1.04892^t)$ in Y1. The percentage change in the population density between 1960 and 2000 is given by the formula $\frac{p(100) - p(60)}{p(60)} \cdot 100\% = \frac{Y1(100) - Y1(60)}{Y1(60)} \cdot 100\%$. Recall that if you type in the entire quotient, you avoid rounding errors.	`(Y₁(100)-Y₁(60))` `/Y₁(60)` `5.756077944` `Ans*100` `575.6077944`

The population density of Nevada increased by about 575.6% between 1960 and 2000.

3.2 Instantaneous Rates of Change

We first examine the principle of local linearity, which says that if you are close enough, the tangent line and the curve are indistinguishable. We also explore two methods for using the calculator to draw a tangent line at a point on a curve.

3.2.1a MAGNIFYING A PORTION OF A GRAPH   The ZOOM menu of your calculator allows you to magnify any portion of a graph. Consider the graph shown in Figure 3.7 in Section 3.2 of *Calculus Concepts*. The temperature model is $T(x) = -0.804x^2 + 11.644x + 38.114$ degrees Fahrenheit where x is the number of hours after 6 a.m.

Enter the temperature equation in Y1 and draw the graph between 6 a.m. ($x = 0$) and 6 p.m. ($x = 12$) using ZOOM ▲ [ZoomFit]. We now want to zoom and "box" in several points on the graph to see a magnified view at those points.	
The first point we consider on the graph is point A with $x = 3$. Press ZOOM 1 [ZBox] and use the arrow keys (◀, ▲, etc.) to move the cursor to the left of the curve close to where $x = 3$. (You may not have the same coordinates as those shown on the right.) Press ENTER to fix the lower left corner of the box.	`X=2.5531915  Y=65.312586`
Use the arrow keys to move the cursor to the opposite corner of your "zoom" box. Point A should be close to the center of your box.	`X=3.8297872  Y=69.392374`
Press ENTER to magnify the portion of the graph that is inside the box. Look at the dimensions of the view you now see with WINDOW. Repeat the above process if necessary. The closer you zoom in on the graph, the more it looks like a line.	

Reset Xmin to 0, Xmax to 12, and redraw the graph of T with ZoomFit. Repeat the zoom and box steps for point B with $x = 7.24$ and point C with $x = 9$. Note for each point how the graph of the parabola looks more like a line the more you magnify the view. These are illustrations of *local linearity*.

3.2.1b DRAWING A TANGENT LINE The DRAW menu of the TI-83 contains the instruction to draw a tangent line to the graph of a function at a point. To illustrate the process, we draw several tangent lines on the graph of *T*, the temperature function given in the previous section of this *Guide*.

Enter $T(x) = -0.804x^2 + 11.644x + 38.114$ in Y1. Next, press WINDOW and set Xmin = 0, Xmax = 12, Ymin = 38, and Ymax = 85. Press GRAPH. With the graph on the screen, press 2nd PRGM (DRAW) 5 [Tangent(], then press 3 ENTER, and the tangent line is drawn at *x* = 3.	

Note that the equation of the tangent line and the input value of the point of tangency are displayed at the bottom of the screen. Compare this line with the zoomed-in view of the curve at point *A* (in the previous section). They actually are the same line if you are at point *A*!

We illustrate a slightly different method to draw the tangent line to the curve at point *B* where *x* = 7.24. Return to the home screen and press 2nd PRGM (DRAW) 5 [Tangent(] VARS ▶ [Y–VARS] 1 [Function] 1 [Y1] , 7.24)].	
Press ENTER and the tangent line is drawn at the high point of the parabola. Compare this line with the zoomed-in view of the curve at point *B* (in the previous section). Note that this method of drawing the tangent line shows neither the equation of the tangent nor the point of tangency.	
Choose the method you prefer from the two methods given above and draw the tangent at point *C* where *x* = 9. (Compare your graph with the one shown in Figure 3.9 in Section 3.2 of the text.)	

3.2.2 VISUALIZING THE LIMITING PROCESS This section of the *Guide* is optional, but it might help you understand what it means for the tangent line to be the limiting position of secant lines. Program SECTAN is used to view secant lines between a point $(a, f(a))$ and close points on a curve $y = f(x)$. The program also draws the tangent line at the point $(a, f(a))$. We illustrate with the graph of the function *T* that was used in the previous two sections, but you can use this program with any graph. (Program SECTAN is in the *TI-83 Program Appendix*.)

Caution: Before using program SECTAN, a function (using *x* as the input variable) must be entered in the Y1 location of the Y= list. In order to properly view the secant lines and the tangent line, you must first draw a graph clearly showing the function, the point of tangency (which should be near the center of the graph), and a large enough window so that the close points on either side of the point of tangency can be viewed.

Have $T(x) = -0.804x^2 + 11.644x + 38.114$ in Y1. Next, press WINDOW and set Xmin = 0, Xmax = 7, Ymin = 30, and Ymax = 90. Press GRAPH. Press PRGM and the number or letter next to the location of program SECTAN. (Your program list may not look like the one shown on the right.)	

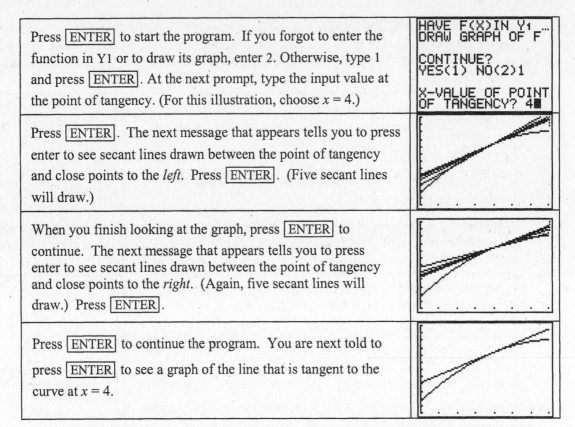

Press ENTER to start the program. If you forgot to enter the function in Y1 or to draw its graph, enter 2. Otherwise, type 1 and press ENTER. At the next prompt, type the input value at the point of tangency. (For this illustration, choose $x = 4$.)	HAVE F(X)IN Y1 ... DRAW GRAPH OF F CONTINUE? YES(1) NO(2)1 X-VALUE OF POINT OF TANGENCY? 4■
Press ENTER. The next message that appears tells you to press enter to see secant lines drawn between the point of tangency and close points to the *left*. Press ENTER. (Five secant lines will draw.)	
When you finish looking at the graph, press ENTER to continue. The next message that appears tells you to press enter to see secant lines drawn between the point of tangency and close points to the *right*. (Again, five secant lines will draw.) Press ENTER.	
Press ENTER to continue the program. You are next told to press ENTER to see a graph of the line that is tangent to the curve at $x = 4$.	

As you watch the graphs, you should notice that the secant lines are becoming closer and closer to the tangent line as the close point moves closer and closer to the point of tangency.

3.2.3 TANGENT LINES AND INSTANTANEOUS RATES OF CHANGE   Sometimes the TI-83 gives results that are not the same as the mathematical results you expect. This does not mean that the calculator is incorrect – it does, however, mean that the calculator programming is using a different formula or definition than the one that you are using. You need to know when the TI-83 produces a different result from what you expect from our formulas.

In this section, we investigate what the calculator does if you ask it to draw a tangent line where the line cannot be drawn. Consider these special cases:

1. What happens if the tangent line is vertical? We consider the function $f(x) = (x + 1)^{1/3}$ which has a vertical tangent at $x = {}^-1$.

2. How does the calculator respond when the instantaneous rate of change at a point does not exist? We illustrate with $g(x) = |x| - 1$, a function that has a sharp point at $(0, {}^-1)$.

3. Does the calculator draw the tangent line at the break point(s) of a piecewise continuous function? We consider two situations:
    a. $y = h(x)$, a piecewise continuous function that is continuous at all points, and
    b. $y = m(x)$, a piecewise continuous function that is not continuous at $x = 1$.

1. Enter the function $f(x) = (x + 1)^{1/3}$ in the Y1 location of the Y= list. Remember that anytime there is more than one symbol in an exponent and you are not sure of the TI-83's order of operations, enclose the power in parentheses.	Plot1 Plot2 Plot3 \Y1■(X+1)^(1/3) \Y2= \Y3= \Y4= \Y5= \Y6= \Y7=

Draw the graph of the function with [ZOOM] 4 [ZDecimal].
With the graph on the screen, press [2nd] [PRGM] (DRAW)
5 [Tangent(], then press [(-)] 1 [ENTER]. The tangent line is
drawn at $x = -1$.

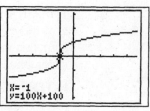

CAUTION: Note that the TI-83 correctly draws the vertical tangent. However, the
equation of the vertical tangent line is $x = -1$, <u>not</u> $y = 100x + 100$! The line whose equation is printed on the calculator screen has slope 100, but the instantaneous rate of change at $(-1, 0)$ does not exist because the tangent is vertical.

2. Clear Y1 and enter the function $g(x) = |x| - 1$. The absolute
   value symbol is typed with [MATH] [▶] 1 [abs(].

Draw the graph of the function with [ZOOM] 4 [ZDecimal].
With the graph on the screen, press [2nd] [PRGM] (DRAW)
5 [Tangent(], then press [ENTER]. The tangent line is drawn
at $x = 0$.

THIS IS INCORRECT! There is a sharp point on the graph of g at $(0, -1)$, and the
limiting positions of secant lines from the left and the right at that point are different. A
tangent line cannot be drawn on the graph of g at $(0, -1)$ according to *our* definition of
instantaneous rate of change. The TI-83's definition is different, and this is why the
line is drawn. (The TI-83 definition is explained in Section 3.5.1 of this *Guide*.)

CAUTION: Note that the TI-83 gives the correct equation for the line that is drawn.
However, <u>there is no tangent line at this sharp point</u> on this continuous function.

3a. Clear Y1 and Y2 and enter, as indicated, the function
$$h(x) = \begin{cases} x^2 & \text{when } x \le 1 \\ x & \text{when } x > 1 \end{cases}$$
. (If you prefer, h can be entered as
a single statement in Y3, as indicated in Section 1.2.3 of this
Guide.) Recall that the inequality symbols are accessed with
[2nd] [MATH] (TEST).

Set each part of the function to draw in DOT mode by placing the cursor over the equals
sign in each function, pressing [◀], and then pressing [ENTER] 6 times until the slanted
line turns to a dotted line.

Draw the graph of the function with [ZOOM] 4 [ZDecimal].
With the graph on the screen, press [2nd] [PRGM] (DRAW)
5 [Tangent(], then press 1 and [ENTER]. The tangent line is
drawn at the break point, $x = 1$.

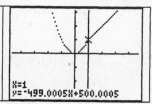

THE GRAPH YOU SEE IS INCORRECT!   Even though h is continuous for all values of x, there is no tangent line to the graph of h at $x = 1$. Secant lines drawn using close points on the right and on the left of $x = 1$ do not approach the same slope, so the instantaneous rate of change of h does not exist at $x = 1$.

NOTE: There is no tangent line at the break point of a piecewise continuous function (even if that function is continuous at the break point) unless secant lines drawn through close points to the left and right of that point approach the same value.

3b. Edit Y2 by placing the cursor over the right parenthesis following X and pressing [2nd] [DEL] (INS) [+] 1 to enter

$$m(x) = \begin{cases} x^2 & \text{when } x \le 1 \\ x+1 & \text{when } x > 1 \end{cases}$$

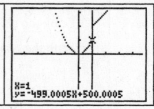

The TI-83 should still be in DOT mode from the previous graph.

Draw the graph of the function with [ZOOM] 4 [ZDecimal].

With the graph on the screen, press [2nd] [PRGM] (DRAW)

5 [Tangent(], then press 1 and [ENTER]. The tangent line is drawn at the break point, $x = 1$.

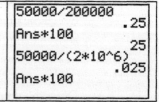

CAUTION: THE GRAPH YOU SEE IS INCORRECT! Because m is not continuous when $x = 1$, the instantaneous rate of change does not exist at that point. The tangent line cannot be drawn on the graph of m when $x = 1$!

Be certain that the instantaneous rate of change exists at a point and the tangent line exists at that point before using your calculator to draw a tangent line or use the tangent equation that may be printed on the screen. Because of the method that your calculator uses to compute instantaneous rates of change, it might draw a tangent line at a point on a curve where the tangent line, according to our definition, does not exist.

 ## 3.3 Derivatives

There are no new calculator techniques in this section, but we illustrate a new calculation.

3.3.1 CALCULATING PERCENTAGE RATE OF CHANGE   Suppose the growth rate of a population is 50,000 people per year and the current population size is 200,000 people.

What is the percentage rate of change of the population? The answer is 25% per year.

Suppose instead that the current population size is 2 million. What is the percentage rate of change? The answer is 2.5% per year, which is a much smaller percentage rate of change.

 ## 3.4 Numerically Finding Slopes

Using your calculator to find slopes of tangent lines does not involve a new procedure. However, the techniques that are discussed in this section allow you to repeatedly apply a method of finding slopes that gives quick and accurate results.

3.4.1a NUMERICALLY ESTIMATING SLOPES ON THE HOME SCREEN   Finding the slopes of secant lines joining the point at which the tangent line is drawn to increasingly close points on a function to the left and right of the point of tangency is easily done using your TI-83. Suppose we want to numerically estimate the slope of the tangent line at $t = 8$ to the graph of the function that gives the number of polio cases in 1949: $y = \dfrac{42{,}183.911}{1 + 21{,}484.253e^{-1.248911t}}$ where $t = 1$ on January 31, 1949, $t = 2$ on February 28, 1949, etc. (See Example 1 in the text.)

Enter the polio cases equation in the Y1 location of the Y= list. (Carefully check your entry of the equation, especially the location of the parentheses.)
 	
 We now evaluate the slopes of secant lines that join close points to the left of $x = 8$ with $x = 8$.
	`Plot1 Plot2 Plot3`
	`\Y1░42183.911/(1`
	`+21484.253e^(-1.`
	`248911X))`
	`\Y2=`
	`\Y3=`
	`\Y4=`
	`\Y5=`

On the home screen, type in the expression shown to the right to compute the slope of the secant line joining the close point where $x = 7.9$ and the point of tangency where $x = 8$.
 	
 Record on paper each slope, to at least 1 more decimal place than the desired accuracy, as it is computed. You are asked to find the nearest whole number that these slopes are approaching, so record at least one decimal place in your table of slopes.
	`(Y1(8)-Y1(7.9))/`
	`(8-7.9)`
	`          13159.68272`

Press [2nd] [ENTER] (ENTRY) to recall the last entry, and then use the arrow keys to move the cursor over the 9 in the "7.9". Press [2nd] [DEL] (INS) and press 9 to insert another 9 in <u>both</u> places that 7.9 appears. Press [ENTER] to find the slope of the secant line joining $x = 7.99$ and $x = 8$.	`(Y1(8)-Y1(7.9))/`
	`(8-7.9)`
	`          13159.68272`
	`(Y1(8)-Y1(7.99))`
	`/(8-7.99)`
	`          13170.61766`

Continue in this manner, recording each result on paper, until you can determine to which value the slopes from the left seem to be getting closer and closer.
 	
 It appears that the slopes of the secant lines from the left are approaching 13,170 cases per month.
	`          13170.18713`
	`(Y1(8)-Y1(7.9999`
	`))/(8-7.9999)`
	`          13170.12882`
	`(Y1(8)-Y1(7.9999`
	`9))/(8-7.99999)`
	`          13170.1229`

We now evaluate the slopes joining the point of tangency and nearby close points to the *right* of $x = 8$.
 	
 Clear the screen, recall the last expression with [2nd] [ENTER] (ENTRY), and edit it so that the nearby point is $x = 8.1$. Press [ENTER] to calculate the secant line slope.
	`(Y1(8)-Y1(8.1))/`
	`(8-8.1)`
	`          13146.38421`

Continue in this manner as you did when calculating slopes to the left, but each time insert a 0 before the "1" in two places in the close point. Record each result on paper until you can determine the value the slopes from the right are approaching.
 	
 It appears that the slopes of the secant lines from the right are approaching 13,170 cases per month.
	`          13170.0538`
	`(Y1(8)-Y1(8.0001`
	`))/(8-8.0001)`
	`          13170.11549`
	`(Y1(8)-Y1(8.0000`
	`1))/(8-8.00001)`
	`          13170.1214`

When the slopes from the left and the slopes from the right approach the same number, that number is the slope of the tangent line at the point of tangency. In this case, we estimate the slope of the tangent line to be 13,170 cases per month.

3.4.1b NUMERICALLY ESTIMATING SLOPES USING THE TABLE   The process discussed in Section 3.4.1a of this *Guide* can be done in fewer steps and with fewer keystrokes when you use the TI-83 TABLE. The point of tangency is $x = 8$, $y = Y1(8)$, and let's call the close point $(x, Y1(x))$. Then, slope $= \dfrac{\text{rise}}{\text{run}} = \dfrac{Y1(X) - Y1(8)}{X - 8}$. We illustrate numerically estimating the slope using the TABLE with the logistic function given in Section 3.4.1a of this *Guide*.

Have the polio cases equation given in the previous section in the Y1 location of the Y= list. Enter the above slope formula in another location, say Y2. (Also remember to enclose both the numerator and denominator of the slope formula in parentheses.) Turn off Y1 because we are only considering the output from Y2.	```Plot1 Plot2 Plot3```
	 `\Y1=42183.911/(1`
	 `+21484.253e^(-1.`
	 `248911X))`
	 `\Y2=(Y1(X)-Y1(8)`
	 `)/(X-8)`
	 `\Y3=`
	 `\Y4=`

Press 2nd WINDOW (TBLSET) and choose ASK in the Indpnt: location. (See page A-8 for more specific instructions.) Access the table with 2nd GRAPH (TABLE) and delete or type over any previous entries in the X column. Enter values for X, the input of the close point, so that X gets closer and closer to 8 from the left.	X | Y2 7.9 | 13160 7.99 | 13171 7.999 | 13170 7.9999 | 13170 8 | 13170 `X=7.99999`

- Notice that as you continue to enter numbers, the calculator switches the input values you enter to scientific notation and displays rounded output values so that the numbers can fit on the screen in the space allotted for outputs of the table. You should position the cursor over each output value and record on paper as many decimal places as are necessary to determine the limit from the left to the desired degree of accuracy.

Repeat the process, entering values for X, the input of the close point, so that X gets closer and closer to 8 from the right. Remember to highlight the output values and record on paper the slope to at least one more decimal position than the desired accuracy.	X | Y2 8.1 | 13146 8.01 | 13169 8.001 | 13170 8.0001 | 13170 8 | 13170 `X=8.00001`
The limit is estimated to be the value that the limits from the left and the right are approaching: 13,170 cases per month. *Note:* You may wish to leave the slope formula in Y2 as long as you need it. Turn Y2 off when you are not using it.	X | Y2 8.1 | 13146 8.01 | 13169 8.001 | 13170 8.0001 | 13170 8 | 13170 `Y2=13170.1214`

3.5 Algebraically Finding Slopes

The TI-83 does not find algebraic formulas for slope, but you can use the built-in numerical derivative and draw the graph of a derivative to check any formula that you find algebraically.

3.5.1 UNDERSTANDING YOUR CALCULATOR'S SLOPE FUNCTION   The TI-83 uses the slope of a secant line to approximate the slope of the tangent line at a point on the graph of a function. However, instead of using a secant line through the point of tangency and a close point, the TI-83 uses the slope of a secant line through two close points that are equally spaced from the point of tangency.

The secant line joining the points $(a - k, f(a - k))$ and $(a + k, f(a + k))$ and the line tangent to the graph of f where $x = a$ are shown in Figure 4. Notice that the slopes of the secant line and tangent appear to be close to the same value even though these are different lines.

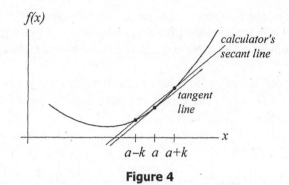

Figure 4

As k gets closer and closer to 0, the two points through which the secant line passes move closer and closer to a. Provided the slope of the tangent line exists, the limiting position of the secant line will be the tangent line.

The calculator's notation for the slope of the secant line shown in Figure 4 is

$$\text{nDeriv(function, symbol for input variable, } a, k)$$

Specifying the value of k is optional. If a value for k is not given, the calculator automatically uses $k = 0.001$. Any smooth, continuous function will do, so let's investigate these ideas with the function in Example 3 of Section 3.5 of *Calculus Concepts*: $f(x) = 2\sqrt{x}$.

Enter $f(x) = 2\sqrt{x}$ in one of the locations of the Y= list, say Y1. Return to the home screen with 2nd MODE (QUIT).	```Plot1 Plot2 Plot3 \Y1■2√(X) \Y2= \Y3= \Y4= \Y5= \Y6= \Y7=```
Suppose you want to find the slope of the secant line between the points $(0, f(0))$ and $(2, f(2))$. That is, you are finding the slope of the secant line between the points $(a - k, f(a - k))$ and $(a + k, f(a + k))$ for $a = 1$ and $k = 1$. Type the expression on the right and then press ENTER. Access nDeriv(with MATH 8 [nDeriv(].	```nDeriv(Y1,X,1,1) 1.414213562```

Remember that when you are on the home screen, you can recall previous instructions with the keystrokes 2nd ENTER (ENTRY).

Recall the last entry and edit the expression so that k changes from 1 to 0.1. Press ENTER. Again recall the last entry, and edit the expression so that k changes from 0.1 to 0.01. Press ENTER.	``` 1.414213562
	nDeriv(Y1,X,1,.1
	)
	        1.001255501
	nDeriv(Y1,X,1,.0
	1)
	        1.000012501```

Repeat the process for $k = 0.001$ and $k = 0.0001$. Note how the results are changing. As k becomes smaller and smaller, the secant line slope is becoming closer and closer to 1.	``` 1.000012501
	nDeriv(Y1,X,1,.0
	01)
	        1.000000125
	nDeriv(Y1,X,1,.0
	001)
	        1.000000002```

- A logical conclusion is that the tangent line slope is 1. However, realize that we have just done another type of numerical investigation, not an algebraic proof.

In the table below, the first row lists some values of a, the input of a point of tangency, and the second row gives the actual slope (to 7 decimal places) of the tangent line at those values. The algebraic method gives the exact slope of the line tangent to the graph of f at these input values.

Use your calculator to verify the values in the third through sixth rows that give the slope of the secant line (to 7 decimal places) between the points $(a - k, f(a - k))$ and $(a + k, f(a + k))$ for the indicated values of k. Find each secant line slope with nDeriv(Y1, X, a, k).

a = input of point of tangency	2.3	5	12.82	62.7
slope of tangent line = $f'(a)$	0.6593805	0.4472136	0.2792904	0.1262892
slope of secant line, $k = 0.1$	0.6595364	0.4472360	0.2792925	0.1262892
slope of secant line, $k = 0.01$	0.6593820	0.4472138	0.2792904	0.1262892
slope of secant line, $k = 0.001$	0.6593805	0.4472136	0.2792904	0.1262892
slope of secant line, $k = 0.0001$	0.6593805	0.4472136	0.2792904	0.1262892

You can see that the slope of the secant line is very close to the slope of the tangent line for small values of k. For this function, the slope of the secant line does a great job of approximating the slope of the tangent line when k is very small.

Now repeat the process, but do not include k in the instruction. That is, find the secant line slope by calculating nDeriv(Y1, X, a, k). Did you obtain the following (to 7 decimal places)?

slope of secant line	0.6593805	0.4472136	0.2792904	0.1262892

These values are those in the fifth row of the above table – the values for $k = 0.001$. From this point forward, we use $k = 0.001$ and therefore do not specify k when evaluating nDeriv(. Will the slope of this secant line always do a good job of approximating the slope of the tangent line when $k = 0.001$? Yes, it generally does, as long as the instantaneous rate of change exists at the input value at which you evaluate nDeriv(. When the instantaneous rate of change does not exist, nDeriv(should *not* be used to approximate something that does not have a value!

You will benefit from reading again Section 3.2.3 of this *Guide*, which illustrates several cases when the TI-83's slope function gives a value for the slope when it does not exist. For instance, the instantaneous rate of change of $f(x) = |x|$ does not exist at $x = 0$ because the graph of f has a sharp point there. However, the TI-83's slope function does exist at that point!	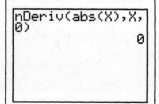

Chapter 4 Determining Change: Derivatives

4.3 More Simple Rate-of-Change Formulas

The TI-83 only approximates numerical values of slopes – it does not give a slope in formula form. You also need to remember that the TI-83 calculates the slope (*i.e.*, the derivative) at a specific input value by a different method than the one we use to calculate the slope.

4.3.1 DERIVATIVE NOTATION AND CALCULATOR NOTATION

In addition to learning when your calculator gives an acceptable answer for a derivative and when it does not, you also need to understand the differences and similarities in mathematical derivative notation and calculator notation. The notation that is used for the calculator's numerical derivative is nDeriv($f(x), x, x$). The correspondence between the mathematical derivative notation $\frac{df(x)}{dx}$ and the calculator's notation nDeriv($f(x), x, x$) is illustrated in Figure 5.

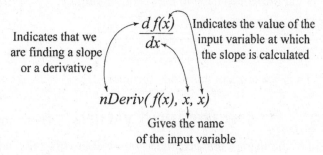

Figure 5

We illustrate another use of the TI-83's derivative by constructing Table 4.7 in Section 4.3 of *Calculus Concepts*. The table lists x, $y = f(x) = e^x$, and $y' = \frac{df}{dx}$ for 4 different inputs. We next evaluate this function at these 4 and several other inputs.

You can evaluate the TI-83 numerical derivative on the home screen or in the table. We choose to use the table. Enter the function f in the Y= list, say in Y1. In Y2 enter the numerical derivative evaluated at a general input X.	Plot1 Plot2 Plot3 \Y1◘e^(X) \Y2◘nDeriv(Y1,X, X) \Y3= \Y4= \Y5= \Y6=
Press [2nd] [WINDOW] (TBLSET) and choose ASK in the Indpnt: location. Access the table with [2nd] [GRAPH] (TABLE) and delete or type over any previous entries in the X column. Enter the values for X that are shown on the screen to the right.	X | Y1 | Y2 -3 | .04979 | .04979 -2 | .13534 | .13534 -1 | .36788 | .36788 0 | 1 | 1 1 | 2.7183 | 2.7183 2 | 7.3891 | 7.3891 3 | 20.086 | 20.086 X=3

It appears that the derivative values are the same as the function outputs. In fact, this is a true statement for all inputs of f – this function is its own derivative!

You can use the methods discussed in Section 1.3.2a of this Guide to find the values used in Table 4.9 to numerically estimate $\lim\limits_{h \to 0} \frac{2^h - 1}{h}$. Instead of this, we explore an alternate method of confirming that $\frac{d(2^x)}{dx} = (\ln 2) \, 2^x$.

Press $\boxed{Y=}$ and edit Y1 to be the function $g(x) = 2^x$.

Access the statistical lists, clear any previous entries from L1, L2, L3, and L4. Enter the x-values shown above in L1. Highlight L2 and enter Y1(L1). Remember to type L1 using $\boxed{2nd}$ 1 (L1).

L1	🔲	L3	2
-3	------	------	
-2			
-1			
0			
1			
2			
3			
L2 =Y₁(L₁)			

Press $\boxed{ENTER}$ to fill L2 with the function outputs. Then, highlight L3 and type Y2(L1).

L1	L2	🔲	3
-3	.125	------	
-2	.25		
-1	.5		
0	1		
1	2		
2	4		
3	8		
L3 =Y₂(L₁)			

Press $\boxed{ENTER}$ to fill L3 with the derivative of Y1 evaluated at the inputs in L1. Note that these values are not the same as the function outputs.

To see what relation the slopes have to the function outputs, press $\boxed{\blacktriangleright}$ and highlight L4. Type L3 $\boxed{\div}$ L2 $\boxed{ENTER}$.

L2	L3	L4	4
.125	.08664	.69315	
.25	.17329	.69315	
.5	.34657	.69315	
1	.69315	.69315	
2	1.3863	.69315	
4	2.7726	.69315	
8	5.5452	.69315	
L4(1)=.69314723604			

It appears that the slope values are a multiple of the function output. In fact, that multiple is $\ln 2 \approx 0.693147$. Thus we confirm this slope formula: If $g(x) = 2^x$, then $\frac{dg}{dx} = (\ln 2)\, 2^x$.

4.3.2a CALCULATING $\frac{dy}{dx}$ AT SPECIFIC INPUT VALUES The previous two sections of this *Guide* examined the calculator's numerical derivative nDeriv($f(x)$, x, a) and illustrated that it gives a good approximation to the slope of the tangent line at points where the instantaneous rate of change exists. You can also evaluate the calculator's numerical derivative from the graphics screen using the CALC menu. However, instead of being named nDeriv(in that menu, it is called *dy/dx*. We illustrate this use with the function in part *a* of Example 2 in Section 4.3.

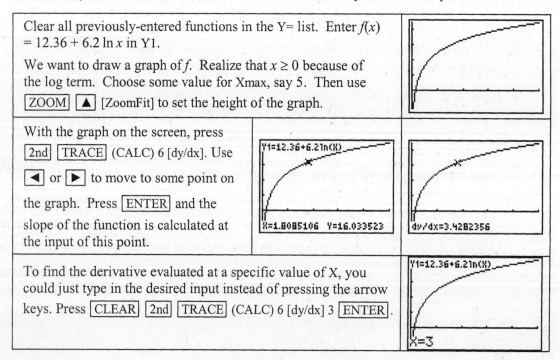

Clear all previously-entered functions in the Y= list. Enter $f(x) = 12.36 + 6.2 \ln x$ in Y1.

We want to draw a graph of f. Realize that $x \geq 0$ because of the log term. Choose some value for Xmax, say 5. Then use $\boxed{ZOOM}$ $\boxed{\blacktriangle}$ [ZoomFit] to set the height of the graph.

With the graph on the screen, press $\boxed{2nd}$ $\boxed{TRACE}$ (CALC) 6 [dy/dx]. Use $\boxed{\blacktriangleleft}$ or $\boxed{\blacktriangleright}$ to move to some point on the graph. Press $\boxed{ENTER}$ and the slope of the function is calculated at the input of this point.

To find the derivative evaluated at a specific value of X, you could just type in the desired input instead of pressing the arrow keys. Press $\boxed{CLEAR}$ $\boxed{2nd}$ $\boxed{TRACE}$ (CALC) 6 [dy/dx] 3 $\boxed{ENTER}$.

> The slope dy/dx = 2.0666667 appears at the bottom of the screen. Return to the home screen and press $\boxed{X,T,\theta,n}$. The TI-83's X memory location has been updated to 3. Now type the numerical derivative instruction (evaluated at 3) as shown to the right. This is the dy/dx value you saw on the graphics screen.

```
X
                    3
nDeriv(Y₁,X,3)
            2.066666743
```

You can use the ideas presented above to check your algebraic formula for the derivative. We next investigate this procedure.

4.3.2b **NUMERICALLY CHECKING SLOPE FORMULAS** It is always a good idea to check your answer. Although your calculator cannot give you an algebraic formula for the derivative function, you can use numerical techniques to check your algebraic derivative formula. The basic idea of the checking process is that if you evaluate your derivative and the TI-83's numerical derivative at several randomly chosen values of the input variable and the outputs are basically the same values, your derivative is *probably* correct.

These same procedures are applicable when you check your results (in the next several sections) after applying the Sum Rule, the Chain Rule, or the Product Rule. We use the function in part *c* of Example 2 in Section 4.3 of *Calculus Concepts* to illustrate.

> Enter $m(r) = \frac{8}{r} - 12\sqrt{r}$ in Y1 (using X as the input variable).
>
> Compute $m'(r)$ using pencil and paper and the derivative rules. Enter this function in Y2. (What you enter in Y2 may or may not be the same as what appears to the right.)
>
> Enter the TI-83's numerical derivative of Y1 (evaluated at a general input X) in Y3. Because you are interested in seeing if the outputs of Y2 and Y3 are the same, turn off Y1.

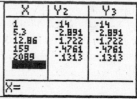

> Press $\boxed{2nd}$ $\boxed{WINDOW}$ (TBLSET) and choose ASK in the Indpnt: location. Access the table with $\boxed{2nd}$ $\boxed{GRAPH}$ (TABLE) and delete or type over any previous entries in the X column. Enter at least three different values for X.

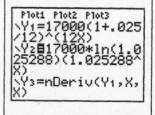

The table gives strong evidence that that Y2 and Y3 are the same function.

4.3.2c **GRAPHICALLY CHECKING SLOPE FORMULAS** When it is used correctly, a graphical check of your algebraic formula works well because you can look at many more inputs when drawing a graph than when viewing specific inputs in a table. We illustrate this use with the function in part *d* of Example 2 in Section 4.3 of *Calculus Concepts*.

> Enter $j(y) = 17{,}000\left(1 + \frac{0.025}{12}\right)^{12y}$ in Y1, using X as the input variable. Next, using pencil and paper and the derivative rules, compute $j'(y)$. Enter this function in Y2.
>
> Enter the TI-83's numerical derivative of Y1 (evaluated at a general input X) in Y3. Before proceeding, turn off the graphs of Y1 and Y3.

```
Plot1 Plot2 Plot3
\Y₁=17000(1+.025
/12)^(12X)
\Y₂=17000*ln(1.0
25288)(1.025288^
X)
\Y₃=nDeriv(Y₁,X,
X)
```

To graphically check your derivative formula answer, you now need to find a good graph of Y2. Because this function is not in a context with a given input interval, the time it takes to find a graph is shortened if you know the approximate shape of the graph. Note that the graph of the function in Y2 is an increasing exponential curve.

Start with ZOOM 4 [ZDecimal] or ZOOM 6 [ZStandard]. Neither of these shows a graph (because of the large coefficient in the function), but you can press TRACE to see some of the output values. Using those values, reset the window. The graph to the right was drawn in the window [−10, 10] by [330, 550], but any view that shows the graph will do.	
Now, press Y= , turn off Y2 and turn on Y3. (Recall that Y3 holds the formula for the derivative of Y1 as computed numerically by the TI-83.) Press GRAPH . Note that you are drawing the graph of Y3 in *exactly the same window* in which you graphed Y2.	

If you see the same graph, your algebraic formula (in Y2) is very likely correct.

4.4 The Chain Rule

You probably noticed that checking your answer for a slope formula graphically is more difficult than checking your answer using the TI-83 table if you have to spend a lot of time finding a window in which to view the graph. Practice with these methods will help you determine which is the best to use to check your answer. (These ideas also apply to Section 4.5.)

4.4.1 SUMMARY OF CHECKING METHODS Before you begin checking your answer, make sure that you have correctly entered the function. It is very frustrating to miss the answer to a problem because you have made an error in entering a function in your calculator. We summarize the methods of checking your algebraic answer using the function in Example 3 of Section 4.4 of *Calculus Concepts*.

Enter the function $P(t) = \dfrac{84.4}{1+33.6\,e^{-0.484t}}$ in the Y1 location, your answer for P' in Y2, and the TI-83 derivative in Y3. Turn off Y1.	Plot1 Plot2 Plot3 \Y1=84.4/(1+33.6 e^(-.484X)) \Y2=1372.546e^(- .484X)/(1+33.6e^ (-.484X))² \Y3=nDeriv(Y1,X, X)
Go to the table, which has been set to ASK mode. Enter at least 3 input values. (Because this problem is in a context, read the problem to see which inputs make sense.) It seems that the answer in Y2 is probably correct!	X · Y₂ · Y₃ 1 · 1.7951 · 1.7951 11 · 4.9389 · 4.9389 20 · .08545 · .08545 X=
If you prefer a graphical check, the problem states that the equation is valid between 1980 and 2001 (and the input is the number of years after 1980). So, turn off Y3, set Xmin = 0, Xmax = 21, and draw a graph of Y2 using ZoomFit. Then, <u>using the same window</u>, draw a graph of Y3 with Y2 turned off. The graphs are the same, again suggesting a correct answer in Y2.	

Chapter 5   Analyzing Change: Applications of Derivatives

 ## 5.2 Relative and Absolute Extreme Points

Your calculator can be very helpful for checking your analytic work when you find optimal points and points of inflection. When you are not required to show work using derivative formulas or when an approximation to the exact answer is all that is required, it is a simple process to use your calculator to find optimal points and inflection points.

5.2.1a FINDING *X*-INTERCEPTS OF SLOPE GRAPHS   Where the graph of a function has a relative maximum or minimum, the slope graph has a horizontal tangent. Where the tangent line is horizontal, the derivative of the function is zero. Thus, finding where the slope graph *crosses* the input axis is the same as finding the input of a relative extreme point.

Consider, for example, the model for Acme Cable Company's revenue for the 26 weeks after it began a sales campaign, where x is the number of weeks since Acme began sales:

$$R(x) = -3x^4 + 160x^3 - 3000x^2 + 24{,}000x \text{ dollars}$$

In Example 2 of Section 5.2 of *Calculus Concepts,* we are first asked to determine when Acme's revenue peaked during the 26-week interval.

Enter R in the Y1 location of the Y= list. Enter either the TI-83's derivative or your derivative in the Y2 location. Turn off Y1.
 	
(If you use your derivative, be sure to use one of the methods at the end of Chapter 4 in this *Guide* to check that your derivative and the TI-83 derivative are the same.)
	`Plot1 Plot2 Plot3`
	`\Y1=-3X^4+160X^3`
	`-3000X²+24000X`
	`\Y2∎nDeriv(Y1,X,`
	`X)`
	`\Y3=`
	`\Y4=`
	`\Y5=`

The statement of the problem indicates that x should be graphed between 1 and 26. Set this horizontal view, and draw the slope graph in Y2 with ZOOM ▲ [ZoomFit]. For a better view to use in this illustration, reset the window values to Ymin = ⁻800 and Ymax = 3000. Redraw the slope graph with GRAPH.	

Find the intercepts of the slope graph using the CALC menu. With the graph on the screen, press 2nd TRACE (CALC) 2 [zero]. Use ▶ to move the cursor near to, but still to the *left* of, the rightmost x-intercept.	`Y2=nDeriv(Y1,X,X)`
	
	`Left Bound?`
	`X=19.085106  Y=906.17454`

Press ENTER to mark the location of the left bound. Use ▶ to move the cursor near to, but to the *right* of, the rightmost x-intercept.	`Y2=nDeriv(Y1,X,X)`
	
	`Right Bound?`
	`X=20.148936  Y=-184.0868`

- Note that the calculator has marked (at the top of the screen) the left bound with a small triangle. The right bound will be similarly marked. The x-intercept must lie between these two bound marks. If you incorrectly mark the interval, you may not get an answer.

Press ENTER to mark the location of the right bound. You are next asked to provide a guess. Any value near the intercept will do. Use ◄ to move the cursor near the intercept.	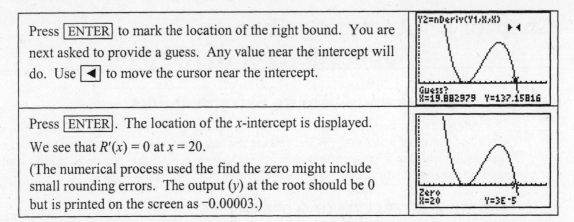

Press ENTER. The location of the x-intercept is displayed. We see that $R'(x) = 0$ at $x = 20$. (The numerical process used the find the zero might include small rounding errors. The output (y) at the root should be 0 but is printed on the screen as ⁻0.00003.)	

As requested in part *b* of Example 2, we now need to determine if the slope graph crosses the x-axis, only touches the x-axis, or does neither at the other location that is an intercept. Note that you can find that $x = 10$ at this point using the same procedure as described above.

Use Zoom In (see Section 1.3.2b) or ZBox (see Section 3.2.1a) as many times as necessary to magnify the portion of the graph around $x = 10$ in order to examine it more closely. We choose to use ZBox, but both work equally well.	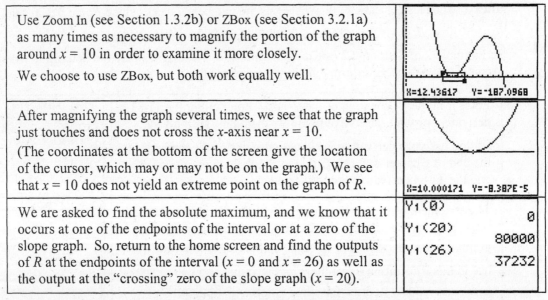

After magnifying the graph several times, we see that the graph just touches and does not cross the x-axis near $x = 10$. (The coordinates at the bottom of the screen give the location of the cursor, which may or may not be on the graph.) We see that $x = 10$ does not yield an extreme point on the graph of R.	

We are asked to find the absolute maximum, and we know that it occurs at one of the endpoints of the interval or at a zero of the slope graph. So, return to the home screen and find the outputs of R at the endpoints of the interval ($x = 0$ and $x = 26$) as well as the output at the "crossing" zero of the slope graph ($x = 20$).	Y₁(0) 0 Y₁(20) 80000 Y₁(26) 37232

We see that Acme's revenue was greatest at 20 weeks after they began the sales campaign.

5.2.1b FINDING ZEROS OF SLOPE FUNCTIONS USING THE SOLVER

You may find it more convenient to use the TI-83 SOLVER rather than the graph and the CALC menu to find the x-intercept(s) of the slope graph. We illustrate using Acme Cable's revenue function.

Enter R in the Y1 location of the Y= list. Enter either the TI-83's derivative or your derivative in the Y2 location. Turn off Y2.	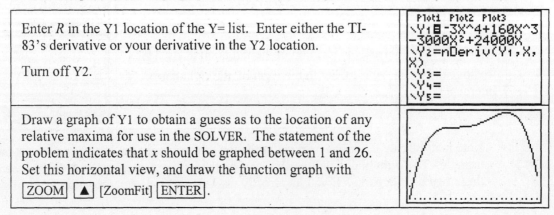

Draw a graph of Y1 to obtain a guess as to the location of any relative maxima for use in the SOLVER. The statement of the problem indicates that x should be graphed between 1 and 26. Set this horizontal view, and draw the function graph with ZOOM ▲ [ZoomFit] ENTER.	

Reset Ymax to a larger value, say 95,000, to better see the high point on the graph. Graph R and press [TRACE]. Hold down [▶] until you have an estimate of the input location of the high point. The maximum seems to occur when x is near 19.9.	Y1=-3X^4+160X^3-3000X2+_ ⟨graph⟩ X=19.882979 . Y=79991.911
Access the SOLVER with [MATH] 0 and press [▲] to go to the eqn: location. Clear what is there from a previous problem and enter Y2. (Recall that we want to solve the equation $R'(x) = 0$ to find where the graph of R has a maximum or minimum.)	EQUATION SOLVER eqn: 0=Y2
Press [ENTER]. Type in the guess that you obtained for the location of the maximum by tracing the graph of R. With the cursor on the line containing X, press [ALPHA] [ENTER] (SOLVE).	Y2=0 X=19.9■ bound=⟨-1E99,1…
The solution $x = 20$ is found. Recall that calculators use numerical algorithms to find zeros. You may or may not obtain the exact value 20. Always round the answer obtained from the SOLVER to make sense in the problem context (here, round to a whole number.)	Y2=0 ▪X=19.999999969… bound=⟨-1E99,1… ▪left-rt=1E-5
Note that if you enter $x \approx 10$ as a guess in the SOLVER, the SOLVER will still return the solution $x = 20$. Acme's revenue was greatest at 20 weeks after they began the sales campaign.	Y2=0 ▪X=19.999999939… bound=⟨-1E99,1… ▪left-rt=-2E-5

5.2.1c USING THE CALCULATOR TO FIND OPTIMAL POINTS Once you draw a graph of a function that clearly shows the optimal points, your calculator can find the location of those high points and low points without using calculus. However, we recommend not relying only on this method because your instructor may ask you to show your work using derivatives. If so, this method would probably earn you no credit! This method does give a good check of your answer, and we illustrate it using Acme Cable's revenue function R from Section 5.2.1a.

Enter R in the Y1 location of the Y= list. The statement of the problem indicates that x should be graphed between 1 and 26. Set this horizontal view, and draw the function graph with [ZOOM] [▲] [ZoomFit] [ENTER]. Press [WINDOW] and reset Ymax to 95,000. Press [GRAPH].	⟨graph⟩
With the graph of R on the screen, press [2nd] [TRACE] (CALC) 4 [maximum]. Use [▶] to move the cursor near, but still to the *left* of, the high point on the curve.	Y1=-3X^4+160X^3-3000X2+_ ⟨graph⟩ LeftBound? X=17.223404 . Y=76908.491

Press ENTER to mark the left bound of the interval. Use ▶ to move the cursor to the *right* of the high point on the curve.	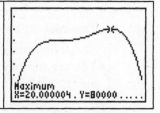
Press ENTER to mark the right bound of the interval. Use ◀ to move the cursor to your guess for the high point on the curve.	

CAUTION: Notice the small arrowheads at the top of the screen that mark the bounds of the interval. The TI-83 returns the highest point that is within the bounded interval that you have marked when you press ENTER. If the relative maximum does not lie in this interval, the TI-83 will return, instead, the highest point in the interval you marked (usually an endpoint).

Press ENTER and the location of the relative maximum (x) and the relative maximum value (y) are displayed. (As previously mentioned, you may not see the exact answer due to rounding errors in the numerical routine used by the TI-83. Always round your answer to make sense in the context.)	

- The method shown in this section also applies to finding the relative minimum values of a function. The only difference is that to find the relative minimum instead of the relative maximum, initially press 2nd TRACE (CALC) 3 [minimum].

 ## 5.3 Inflection Points

As was the case with optimal points, your calculator can be very helpful in checking algebraic work when you find points of inflection. You can also use the methods illustrated in Section 5.2.1c of this *Guide* to find the location of any maximum or minimum points on the graph of the first derivative to find the location of any inflection points for the function.

5.3.1 FINDING *X*-INTERCEPTS OF A SECOND-DERIVATIVE GRAPH   We first look at using the algebraic method of finding inflection points – finding where the graph of the second derivative of a function *crosses* the input axis.

To illustrate, we consider a model for the percentage of students graduating from high school in South Carolina from 1982 through 1990 who entered post-secondary institutions:

$$f(x) = -0.1057x^3 + 1.355x^2 - 3.672x + 50.792 \text{ percent}$$

where x is the number of years after 1982.

Enter f in the Y1 location of the Y= list, the first derivative of f in Y2, and the second derivative of f in Y3. (Be careful not to round any decimal values.) Turn off Y1 and Y2.	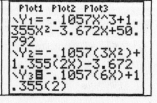

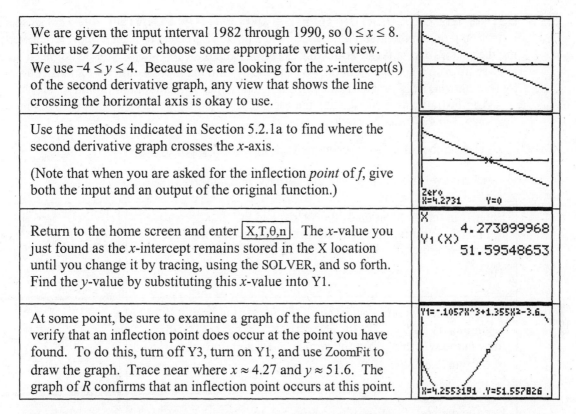

We are given the input interval 1982 through 1990, so $0 \leq x \leq 8$. Either use ZoomFit or choose some appropriate vertical view. We use $-4 \leq y \leq 4$. Because we are looking for the x-intercept(s) of the second derivative graph, any view that shows the line crossing the horizontal axis is okay to use.	
Use the methods indicated in Section 5.2.1a to find where the second derivative graph crosses the x-axis. (Note that when you are asked for the inflection *point* of f, give both the input and an output of the original function.)	
Return to the home screen and enter $\boxed{X,T,\theta,n}$. The x-value you just found as the x-intercept remains stored in the X location until you change it by tracing, using the SOLVER, and so forth. Find the y-value by substituting this x-value into Y1.	
At some point, be sure to examine a graph of the function and verify that an inflection point does occur at the point you have found. To do this, turn off Y3, turn on Y1, and use ZoomFit to draw the graph. Trace near where $x \approx 4.27$ and $y \approx 51.6$. The graph of R confirms that an inflection point occurs at this point.	

EXPLORE: What would the line tangent to the graph of f look like at the inflection point? Use the graph of Y1 and the first method explained in Section 3.2.1b of this *Guide* (with $x = 4.27$) to see if you are correct.

EXPLORE: What is true about the graph of f' at the inflection point? Use the graph of Y2 and the trace cursor to determine if your guess is correct.

- The TI-83 will usually draw an accurate graph of the first derivative of a function when you use nDeriv(. However, this calculator does not have a built-in method to calculate or graph f'', the second derivative. As illustrated below, you can try to use nDeriv(f', X, X) to find f''. Be cautioned, however, that nDeriv(f', X, X) sometimes "breaks down" and gives invalid results. If this should occur, the graph of nDeriv(f', X, X) appears very jagged and this method should not be used.

Enter f in the Y1 location of the Y= list, the first derivative of f in Y2, and the second derivative of f in Y3, using the TI-83's numerical derivative for each derivative that you enter. Turn off Y1. Be sure that Xmin = 0 and Xmax = 8 (as the problem directions indicate.)	
Draw the graph of the first derivative (Y2) and the second derivative (Y3) of f using an appropriate window. You can use ZoomFit to set the vertical view or experiment until you find a suitable view. The graph to the right uses $-4 \leq y \leq 4$.	

Find the x-intercept of the second derivative graph as indicated in this section or find the input of the high point on the first derivative graph (see Section 5.2.1c) to locate the inflection point.

5.3.2 USING THE CALCULATOR TO FIND INFLECTION POINTS   Remember that an inflection point on the graph of a function is a point of greatest or least slope. Whenever finding the second derivative of a function is tedious algebraically and/or you do not need an exact answer from an algebraic solution, you can easily find the input location of an inflection point by finding where the first derivative of the function has a maximum or minimum slope.

We illustrate this method using the logistic function for polio cases that is in Example 2 of Section 5.3 in *Calculus Concepts*:

The number of polio cases in the U.S. in 1949 is given by $C(t) = \dfrac{42{,}183.911}{1 + 21{,}484.253e^{-1.248911t}}$

where $t = 1$ in January, $t = 2$ in February, and so forth.

Enter C in the Y1 location of the Y= list and the first derivative of f in Y2. (You can use your algebraic formula for the first derivative or the calculator's numerical derivative.)
 	
Turn off Y1.
	`Plot1 Plot2 Plot3`
	`\Y1=42183.911/(1`
	`+21484.253e^(-1.`
	`248911X))`
	`\Y2=nDeriv(Y1,X,`
	`X)`
	`\Y3=`
	`\Y4=`

The problem context says that the input interval is from 0 (the beginning of 1949) to 12 (the end of 1949), so set these values for x in the WINDOW. Set the vertical view and draw the graph of C' with ZoomFit.	
Use the methods discussed in Section 5.2.1c ([2nd] [TRACE] (CALC) 4 [maximum]) to find the input location of the maximum point on the slope graph.	`Maximum` `X=7.9870163  Y=13170.986`

The x-value of the maximum of the slope graph is the x-value of the inflection point of the function. To find the rate of change of polio cases at this time, substitute this value of x in Y2. To find the number of cases at this time, substitute x in Y1.	`X`
	`       7.987016258`
	`Y2(X)`
	`        13170.98591`
	`Y1(X)`
	`        21091.92317`

CAUTION: Do not forget to round your answers appropriately (this function should be interpreted discretely) and to give units of measure with each answer.

Note that you could have found the input of the inflection point on the polio cases graph by finding the x-intercept of the second derivative graph. The function that is graphed to the right is Y3 = nDeriv(Y2, X, X), and the graph was drawn using ZoomFit.	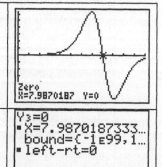

If you prefer, you could have found the input of the inflection point by solving the equation $C'' = 0$ using the SOLVER. (Do not forget that drawing the graph of C and tracing it can be used to find a guess for the SOLVER.)	`Y3=0`
	`•X=7.9870187333...`
	`  bound={-1E99,1...`
	`•left-rt=0`

Chapter 6   Accumulating Change: Limits of Sums and the Definite Integral

 ## 6.1 Results of Change and Area Approximations

So far, we have used the TI-83 to investigate rates of change. In this chapter we consider the second main topic in calculus – the accumulation of change. You calculator has many useful features that will assist in your investigations of the results of change.

6.1.1 AREA APPROXIMATIONS USING LEFT RECTANGLES   The TI-83 lists can be used to approximate, using left rectangles, the area between the horizontal axis and a rate of change function between two specified input values. We illustrate the necessary steps using the data for the number of customers entering a large department store during a Saturday sale. These data appear in Table 6.1 of Section 6.1 in *Calculus Concepts*.

Minutes after 9 a.m.	0	45	75	120	165	195	255	330	370	420	495	570	630	675
Customers per minute	1	2	3	4	4	5	5	5	5	4	4	3	2	2

Enter these data in lists L1 and L2. Find a cubic function to fit the data, and paste the function in Y1. Turn on Plot1.
 	
(If you did not obtain the function shown to the right, check your data values.) We next graph the function on a scatter plot of the data over the 12-hour (720 minute) sale.
	`CubicReg`
	`y=ax`^{3}`+bx`^{2}`+cx+d`
	`a=4.5890356E`$^-$`8`
	`b=-7.781267E`$^-$`5`
	`c=.0330326126`
	`d=.8876300878`

Either set Xmin = 0, Xmax = 720, Xscl = 60 and use [ZOOM] [▲] [ZoomFit] [ENTER] to set the vertical view or press [ZOOM] 9 [ZoomStat] [ENTER]. Set Ymin = 0 and press [GRAPH].	

We want to approximate the area under the cubic graph and above the *x*-axis during the 12-hour period using rectangles of equal width (60 minutes).

Press [STAT] 1 [EDIT], clear lists L1 and L2, and enter the inputs 0, 60, 120, …, 720 in L1 A quick way to do this is to highlight L1 and press [2nd] [STAT] (LIST) [▶] [OPS] 5 [seq(] [X,T,θ,n] [,] [X,T,θ,n] [,] 0 [,] 720 [,] 60 [)]. (Section 2.1.1a gives a more detailed explanation of the sequence command.)	L1 = {0,60,120,18...
Because we are using rectangles with heights determined at the left endpoint of each interval, delete the last value in L1 (720). Substitute each input in L1 into the function in Y1 by highlighting L2 and typing Y1(L1). Press [ENTER].	L2(12) = 1.98725049...

Consider what is now in the lists. L1 contains the left endpoints of the 12 rectangles and L2 contains the heights of the rectangles. If we multiply the heights by the widths of the rectangles (60 minutes) and enter this product in L3, we will have the rectangle areas in L3.

A-61

L1	L2	L3	3
0	.88763	*53.258*	
60	2.5994	155.96	
120	3.8103	228.62	
180	4.58	274.8	
240	4.9678	298.07	
300	5.0333	302	
360	4.8359	290.15	

L3(1)=53.25780526...

Highlight L3 and type 60 [X] [2nd] 2 (L2) [ENTER]. L3 now contains the areas of the 12 rectangles.

All that remains is to add the areas of the 12 rectangles. To do this, return to the home screen and type [2nd] [STAT] (LIST) [▶] [▶] [MATH] 5 [sum] [2nd] 3 (L3) [)] [ENTER].

We estimate that 2574 customers came to the Saturday sale.

```
sum(L3)
          2573.712319
```

6.1.2 AREA APPROXIMATIONS USING RIGHT RECTANGLES
When you use left rectangles to approximate the results of change, the rightmost data point is not the height of a rectangle and is not used in the computation of the left-rectangle area. Similarly, when using right rectangles to approximate the results of change, the leftmost data point is not the height of a rectangle and is not used in the computation of the right-rectangle area. We illustrate the right-rectangle approximation using the function r that is given in Example 2 of Section 6.1.

The rate of change of the concentration of a drug in the bloodstream is modeled by

$$r(t) = \begin{cases} 1.708(0.845^x) & \text{when } 0 \le x \le 20 \\ -10.058 + 2.94 \ln x & \text{when } 20 < x \le 30 \end{cases}$$

where x is the number of days after the drug is first administered.

NOTE: It is not necessary to draw a graph of r, but refer to Section 1.2.3 of this *Guide* if you want to review the instructions for graphing this piecewise continuous function. Place the TI-83 in Dot mode (also discussed in Section 1.2.3) if you intend to draw a graph of r.

Enter the 2 pieces of r in the Y= list as shown on the right. With the cursor in Y3, press [(] [VARS] [▶] 1 [Function] 1 [Y1] [)] [(] [X,T,θ,n] [2nd] [MATH] [TEST] 6 [≤] 20 [)] [+] [(] [VARS] [▶] 1 [Function] 2 [Y2] [)] [(] [X,T,θ,n] [2nd] [MATH] [TEST] 3 [>] 20 [)].

Turn off Y1 and Y2. (See Section 1.2.2b of this *Guide* for instructions on turning functions in the Y= list off and on.)

```
Plot1  Plot2  Plot3
\Y1=1.708(.845^X
)
\Y2=-10.058+2.94
ln(X)
\Y3=(Y1)(X≤20)+(
Y2)(X>20)
\Y4=
```

Part a of Example 2 says to find the change in the drug concentration from $x = 0$ through $x = 20$ using right rectangles of width 2 days. Enter the right endpoints (2, 4, 6, ..., 20) in L1.

Use the piecewise continuous function r in Y3 (or use Y1) to find the rectangle heights. As shown, enter the heights in L2.

L1	L2	L3	2
2	1.2196	------	
4	.87029		
6	.62177		
8	.44396		
10	.317		
12	.22634		
14	.16162		

L2 =Y3(L1)■

- Note that the heights in L2 are the values in Table 6.3 in the text, but the Table 6.3 values have been rounded for printing purposes.

Find the right-rectangle areas by multiplying each entry in L2 by 2. This is the same as multiplying the sum of the heights by 2. On the home screen, press 2 [X] [2nd] [STAT] (LIST) [▶] [▶] [MATH] 5 [sum] [2nd] 2 (L2) [)] [ENTER].

```
2*sum(L2)
          8.235309838
```

Part *b* of Example 2 asks us to use the model and right rectangles of width 2 days to estimate the change in drug concentration from $x = 20$ to $x = 30$. Notice that the signed heights in L4 are the same as those given in the text in Table 6.4 (except for rounding.)

Enter the right endpoints (22, 24, 26, 28, 30) in L3. Use the piecewise continuous function *r* in Y3 (or use Y2) to find the rectangle heights. As shown to the right, enter the signed heights in L4.	![L2 L3 L4 4 table: 1.2196/22/-.9703, .87079/24/-.7145, .62177/26/-.4792, .44396/28/-.2613, .317/30/-.0585, .22634, .16162; L4 =Y₃(L₃)]

Find the signed right-rectangle areas by multiplying each entry in L2 by 2. This is the same as multiplying the sum of the signed heights by 2. On the home screen, press $\boxed{2nd}$ $\boxed{ENTER}$ (ENTRY) to return the sum instruction to the screen, change L2 to L4 with $\boxed{\triangleleft}$ $\boxed{\triangleleft}$ and then type $\boxed{2nd}$ 4 (L4). Press $\boxed{ENTER}$.	2*sum(L2)
	    8.235309838
	2*sum(L4)
	    -4.967703084

6.1.3a AREA APPROXIMATIONS USING MIDPOINT RECTANGLES Areas of midpoint rectangles are found using the same procedures as those used to find left and right rectangle areas except that the midpoint of the base of each rectangle is entered in the first list and no endpoints are deleted. We illustrate the midpoint-rectangle approximation using the function *f* in Example 3 of Section 6.1 in *Calculus Concepts*.

Clear all functions from the Y= list. Next, enter the function $f(x) = \sqrt{1 - x^2}$ in some location of the Y= list, say Y1. (Type the square root symbol with $\boxed{2nd}$ $\boxed{x^2}$.)	Plot1 Plot2 Plot3 \Y1■√(1-X²) \Y2= \Y3= \Y4= \Y5= \Y6= \Y7=
Clear lists L1 and L2. To use 4 midpoint rectangles to approximate the area of the region between the graph of *f* and the *x*-axis between $x = 0$ and $x = 1$, first enter the midpoints of the rectangles (0.125, 0.375, 0.625, 0.875) in L1. As shown to the right, enter the heights of the rectangles in L2.	L1 / L2 / L3 2 .125/.99216 .375/.92702 .625/.78062 .875/.48412 L2 =Y₁(L₁)
You can now multiply each rectangle height by the width 0.25 (entering these values in another list) and sum the rectangle areas or you can multiply the sum of the heights in L2 by 0.25. Either calculation results in the area estimate.	.25*sum(L2) .7959823052

6.1.3b SIMPLIFYING AREA APPROXIMATIONS The procedures used in the previous three sections of this *Guide* can become tedious when the number of rectangles is large. When you have a function $y = f(x)$ in the Y1 location of the Y= list, you will find program NUMINTGL very helpful in determining left-rectangle, right-rectangle, and midpoint-rectangle approximations for accumulated change. Program NUMINTGL (listed in the *TI-83 Program Appendix*) performs automatically all the calculations that you have been doing manually using the lists.

WARNING: This program will not execute properly unless the function that determines the heights uses X as the input variable and is pasted in Y1. If you receive an error message while attempting to run this program, consult Programs in *Troubleshooting the TI-83* in this *Guide*.

We illustrate using this program with the function $f(x) = \sqrt{1-x^2}$ that is used in Example 3 in Section 6.1. This function should be entered in the Y1 location of the Y= list.

Press PRGM followed by the number or letter next to the location of the program and ENTER to start NUMINTGL. (At this point, if you did not enter the function in Y1, press 2 ENTER to exit the program. Enter the function in Y1 and re-run the program). If the function is in Y1, press 1 ENTER to continue.	```ENTER F(X) IN Y1 CONTINUE? YES(1) NO(2) ■```
At the next prompt, pressing 1 ENTER draws the approximating rectangles and pressing 2 ENTER obtains only numerical approximations to the area between the function and the horizontal axis between 0 and 1. For illustration purposes, let's choose to see the rectangles, so press 1 ENTER.	```ENTER F(X) IN Y1 CONTINUE? YES(1) NO(2) 1 DRAW PICTURES? YES(1) NO(2) 1```
At the LEFT ENDPOINT? prompt, press 0 ENTER, and at the RIGHT ENDPOINT? prompt, press 1 ENTER to tell the TI-83 the input interval. You are next shown a menu of choices. Press 4 and ENTER for a midpoint-rectangle approximation.	```ENTER CHOICE: LEFT RECT (1) RIGHT RECT (2) TRAPEZOIDS (3) MIDPT RECT (4) 4```
Input 4 at the N? prompt and press ENTER. A graph of the 4 approximating midpoint rectangles and the function are shown. (Note that the program automatically sets the height of the window based on the left and right endpoints of the input interval.)	
The midpoint-rectangle area is displayed as SUM when you press ENTER. *Note:* All area approximations are displayed with "SUM." You need to remember which method you chose to interpret each result.	```RIGHT RECT (2) TRAPEZOIDS (3) MIDPT RECT (4) 4 N? 4 SUM= .7959823052```
Press ENTER once more and some choices are displayed. Suppose you now want to find the approximating area and see the figure for 10 right rectangles. Press 2 to change the method, and then press 2 ENTER to choose right rectangles.	```ENTER CHOICE 1:CHANGE N 2:CHANGE METHOD 3:QUIT```
Press 10 ENTER at the N? prompt to view the right rectangles. Again press ENTER and the right-rectangle area estimate is displayed.	```RIGHT RECT (2) TRAPEZOIDS (3) MIDPT RECT (4) 2 N? 10 SUM= .7261295816```

• As you gain more experience with approximation methods using rectangles, you probably will not want to choose the program option to draw the approximating rectangles.

> Continue in this manner and find the left-rectangle area or change N and find the left-, right-, or midpoint-rectangle approximations for different numbers of subintervals.
>
> When finished, press ENTER and choose 3 to QUIT the program.

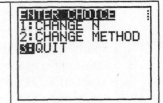

6.2 Limit of Sums, Accumulated Change, and the Definite Integral

This section introduces you to a very important and useful concept of calculus -- the definite integral. Your calculator can be very helpful as you study definite integrals and how they relate to the accumulation of change.

6.2.1 **LIMIT OF SUMS**   When you are looking for a limit in the midpoint-rectangle approximations of the area (or signed area) between a function and the horizontal axis between two values of the input variable, program NUMINTGL is extremely useful! However, when finding a limit of sums using the values displayed by this program, it is not advisable to draw pictures when N, the number of rectangles, is large.

We illustrate using this program to find a limit of sums using $f(x) = \sqrt{1-x^2}$, the function in Example 1 of Section 6.2 of *Calculus Concepts*.

To construct a table of midpoint-rectangle approximations for the area between f and the x-axis from $x = 0$ to $x = 1$, first clear the Y= list, enter f in Y1, and turn off all scatter plots.	`Plot1 Plot2 Plot3`
	`\Y1◼√(1-X²)`
	`\Y2=`
	`\Y3=`
	`\Y4=`
	`\Y5=`
	`\Y6=`
	`\Y7=`

Run program NUMINTGL. At the first prompt, enter 1 to continue and enter 2 to not draw pictures.
 	
At the LEFT ENDPOINT? prompt, enter 0, and at the RIGHT ENDPOINT? prompt, enter 1 to tell the TI-83 that the input interval is from $x = 0$ to $x = 1$.
	`ENTER F(X) IN Y1`
	
	`CONTINUE?`
	`YES(1) NO(2) 1`
	
	`DRAW PICTURES?`
	`YES(1) NO(2) 2`

You are next shown a menu of choices. <u>Always use midpoint rectangles with a limit of sums</u>. Press 4 and ENTER. At the N? prompt, enter 4.	`ENTER CHOICE:`
	`LEFT RECT  (1)`
	`RIGHT RECT (2)`
	`TRAPEZOIDS (3)`
	`MIDPT RECT (4)  4`
	
	`N? 4◼`

Record on paper the area estimate for 4 midpoint rectangles. (You should record at least one more decimal place than the number of places needed for the required accuracy. Because Example 1 asks for 3 decimal-place accuracy, record at least four decimal places of the value shown to the right.)	`RIGHT RECT (2)`
	`TRAPEZOIDS (3)`
	`MIDPT RECT (4)  4`
	
	`N? 4`
	`SUM=`
	`        .7959823052`

Press ENTER to continue the program, and press 1 to choose CHANGE N. Now double the number of rectangles by entering N = 8.
 	
Record the area approximation (again, to four decimal places).
	`N? 8`
	`SUM=`
	`        .7891717328`

Continue in this manner, each time choosing the first option, CHANGE N, and doubling N until a limit is evident.

Intuitively, *finding the limit* means that you are sure what the area approximation, to 3 decimal places (or whatever accuracy is specified in the problem), will be without having to use larger values of N in the program.

When finished, press ENTER and choose 3 to QUIT the program.

 ## 6.4 The Fundamental Theorem

The Fundamental Theorem of Calculus is important because it connects the two main topics in calculus – differentiation and integration. In this section of the *Guide*, we see how to use the TI-83's definite integral function and use it along with the TI-83's numerical derivative to illustrate the Fundamental Theorem in action.

6.4.1a **DEFINITE INTEGRAL NOTATION AND CALCULATOR NOTATION**   Recall that nDeriv is the TI-83's numerical derivative and provides, in most cases, a very good approximation for the instantaneous rate of change of a function when that rate of change exists. As is the case with the numerical derivative, your calculator does not give a formula for accumulation functions. However, it does give an excellent numerical estimate for the definite integral of a function between two specific input values when that integral exists.

The TI's numerical integral is called fnInt, and the correspondence between the mathematical notation $\int_a^b f(x)\,dx$ and the calculator's notation fnInt($f(x)$, x, a, b) is illustrated in Figure 6.

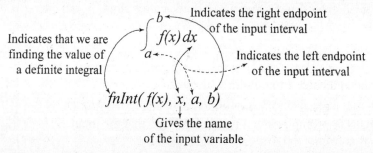

Figure 6

We illustrate the use of the calculator's definite integral function by finding the area from $x = 0$ to $x = 1$ between the function $f(x) = \sqrt{1-x^2}$ and the horizontal axis.

You can enter *f* in Y1 and refer to it in the fnInt expression as Y1. Access fnInt with MATH 9 [fnInt(]. (Any function location can be used. Recall that in the Y= list, you must use X as the input variable name.)	```Plot1 Plot2 Plot3\n\Y1◆√(1-X²)\n\Y2=\n\Y3=\n\Y4=\n\Y5=\n\Y6=\n\Y7=```	```fnInt(Y1,X,0,1)\n .7853984608```
Or, you can enter the function directly on the home screen as shown to the right. When you type a function formula on the home screen, you can use any letter as the input variable symbol.		```fnInt(√(1-X²),X,\n0,1)\n .7853984608\nfnInt(√(1-T²),T,\n0,1)\n .7853984608```

• The TI-83's fnInt function yields the same result (to 3 decimal places) as that found in the limit of sums investigation in Section 6.2.1 of this *Guide*.

6.4.1b THE FUNDAMENTAL THEOREM OF CALCULUS Intuitively, this theorem tells us that the derivative of an antiderivative of a function is the function itself. Let us view this idea both numerically and graphically. The correct syntax for the TI-83's numerical integrator is

fnInt(function, name of input variable, left endpoint for input, right endpoint for input)

Consider the function $f(t) = 3t^2 + 2t - 5$ and the accumulation function $F(x) = \int_1^x f(t)\, dt$.

The Fundamental Theorem of Calculus tells us that $F'(x) = \dfrac{d}{dx}\left(\int_1^x f(t)\, dt\right) = f(x)$; that is, $F'(x)$ is f evaluated at x.

Input f in Y1 and F' in Y2 (remember that the TI-83 requires that you use X as the input variable in the Y= list). Access fnInt with $\boxed{\text{MATH}}$ 9 [fnInt(] and nDeriv with $\boxed{\text{MATH}}$ 8 [nDeriv(]. (Turn off any stat plots that are on.)	Plot1 Plot2 Plot3 \Y1■3X²+2X-5 \Y2■nDeriv(fnInt (Y1,X,1,X),X,X) \Y3= \Y4= \Y5= \Y6=
Have TBLSET set to ASK and press $\boxed{\text{2nd}}$ $\boxed{\text{GRAPH}}$ (TABLE). Input several different values for X. Other than occasional roundoff error because the calculator is approximating these values, the results are identical.	X | Y1 | Y2 -5 | 60 | 60 0 | -5 | -5 4.3 | 59.07 | 59.07 159.82 | 76942 | 76942 1.6667 | 6.6667 | 6.6667 Y1=76941.9372
Find a suitable viewing window such as the one set with $\boxed{\text{ZOOM}}$ 6 [ZStandard]. Without changing the window (that is, draw the graphs by pressing $\boxed{\text{GRAPH}}$), turn off Y2 and draw the graph of Y1. Then turn off Y1 and draw the graph of Y2. (*Note:* The graph of Y2 takes a while to draw.) Turn both Y1 and Y2 on and draw the graph of both functions. Only one graph is seen in each case.	

EXPLORE: Enter several other functions in Y1 and do not change Y2 except possibly for the left endpoint 1 in the fnInt expression. Perform the same explorations as above. Confirm your results with derivative and integral formulas.

6.4.2 DRAWING ANTIDERIVATIVE GRAPHS Recall when using fnInt($f(x)$, x, a, b) that a and b are, respectively, the lower and upper endpoints of the input interval. Also remember that you do not have to use x as the input variable unless you are graphing the integral or evaluating it using the calculator's table.

Unlike when graphing using nDeriv, the TI-83 will not graph a general antiderivative; it only draws the graph of a specific accumulation function. Thus, we can use x for the input at the upper endpoint when we want to draw an antiderivative graph, but not for the inputs at both the upper and lower endpoints.

All of the antiderivatives of a specific function differ only by a constant. We explore this idea using the function $f(x) = 3x^2 - 1$ and its general antiderivative $F(x) = x^3 - x + C$. Because we are working with a general antiderivative in this illustration, we do not have a starting point for the accumulation. We therefore choose some value, say 0, to use as the starting point for the accumulation function to illustrate drawing antiderivative graphs. If you choose a different lower limit, your results will differ from those shown below by a constant.

Enter f in Y1, fnInt(Y1, X, 0, X) in Y2, and F in Y3, Y4, Y5, and Y6, using a different number for C in each function location. (You can use the values of C shown to the right or different values.)	Plot1 Plot2 Plot3 \Y1■3X²−1 \Y2■fnInt(Y1,X,0 ,X) \Y3■X^3−X+1 \Y4■X^3−X−2 \Y5■X^3−X+5 \Y6■X^3−X−4.8
Find a suitable viewing window and graph the functions Y1 through Y6. The graph to the right was drawn with $-3 \le x \le 3$ and $-20 \le y \le 20$.	
It appears that the only difference in the graphs of Y2 through Y6 is the y-axis intercept. But, isn't C the y-axis intercept of each of these antiderivative graphs? Clear Y4, Y5, and Y6. Turn off Y1 and change the 1 in Y3 to 0.	Plot1 Plot2 Plot3 \Y1=3X²−1 \Y2■fnInt(Y1,X,0 ,X) \Y3■X^3−X+0 \Y4= \Y5= \Y6=
Press ⎡GRAPH⎤ and draw the graphs of Y2 and Y3. You should see only one graph. Set the TI-83 TABLE to ASK and enter some values for x. It appears that Y2 and Y3 are the same function.	X Y2 Y3 -5 -120 -120 0 0 0 4.3 75.207 75.207 -8.35 -573.8 -573.8 21 9240 9240 51.2 134167 134167 X=

CAUTION: The methods for checking derivative formulas that were discussed in Sections 4.3.2b and 4.3.2c are not valid for checking general antiderivative formulas. Why not? Because to graph an antiderivative using fnInt, you must arbitrarily choose values for the constant of integration and for the input of the lower endpoint. However, for most of the rate-of-change functions where $f(0) = 0$, the calculator's numerical integrator values and your antiderivative formula values should differ by the same constant at every input value where they are defined.

6.5 The Definite Integral

When using the numerical integrator on the home screen, enter fnInt($f(x)$, x, a, b) for a specific function f with input x and specific values of a and b. (Remember that the input variable does not have to be x when the function formula is entered on the home screen.) If you prefer, f can be in the Y= list and referred to as Y1 (or whatever location is chosen) when using fnInt.

6.5.1a EVALUATING A DEFINITE INTEGRAL ON THE HOME SCREEN   We illustrate the use of fnInt with the function that models the rate of change of the average sea level. The rate-of-change data are given in Table 6.18 of Example 3 in Section 6.5 of *Calculus Concepts*.

Time (thousands of years before the present)	−7	−6	−5	−4	−3	−2	−1
Rate of change of average sea level (meters/year)	3.8	2.6	1.0	0.1	−0.6	−0.9	−1.0

Enter the time values in L1 and the rate of change of the average sea level values in L2. A scatter plot of the data indicates a quadratic function. Find the function and paste it in Y1. (Draw the function on the scatter plot of the data to confirm that it gives a good fit.)	

Because part *a* of Example 3 asks for the *areas* of the regions above and below the input axis and the function, we must find where the function crosses the axis.

You can find this value using the solver (solve Y1 = 0) or by using the graph and the *x*-intercept method that is described in Section 1.1.1i of this *Guide*.

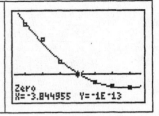

CAUTION: If you round the *x*-intercept value, it will cause whatever you do with this value to not give as an accurate result as possible. This situation occurs many times in this and the next several chapters of the text. Also recall that one of the numerical considerations given in Chapter 1 of *Calculus Concepts* is that intermediate calculation values should not be rounded. For maximum accuracy, store this value in some memory location, say Z, and refer to it as Z in all subsequent calculations.

Whether you use the SOLVER or the *x*-intercept method shown on the calculator screen above, the TI-83 stores the zero in X. Return to the home screen and store the value in Z (or whatever memory location, except X, that you choose). Do not use the X location because the X value changes whenever you use the TI-83 table, the solver, or trace a graph. Type Z with $\boxed{\text{ALPHA}}$ 2.	```X→Z``` ``` -3.844955338```

Find the area of the region above the horizontal axis by typing and entering the first expression shown on the right.
 	
 Find the area of the region below the horizontal axis by typing and entering the second expression shown on the right. (The negative is used because the region is below the input axis.)
	```fnInt(Y1,X,-7,Z)```
	 ``` 5.405934876```
	 ```-fnInt(Y1,X,Z,0)```
	 ``` 2.936490432```

Part *b* of Example 3 asks you to evaluate $\int_{-7}^{0} Y1\,dx$. Find this value by typing and entering the expression shown on the right. Note that the result is not the sum of the two areas – it is their difference.	```fnInt(Y1,X,-7,0)``` ``` 2.469444444```

- If you evaluate a definite integral using antiderivative formulas and check your answer with the calculator using fnInt, you might find a slight difference in the last few decimal places. Remember that the TI-83 is evaluating the definite integral using an approximation technique, not an algebraic formula.

6.5.1b EVALUATING A DEFINITE INTEGRAL FROM THE GRAPHICS SCREEN   The value of a definite integral can also be found from the graphics screen. We again illustrate the use of fnInt with the function that models the rate of change of the average sea level.

Turn off Plot 1 and have the function modeling the average sea level in Y1. (See Section 6.5.1a of this *Guide*.) We want to evaluate $\int_{-7}^{0} Y1\,dx$. (You can practice using this graphical method by finding the areas of the regions indicated in part *c* of Example 4 in Section 6.5 of the text.)	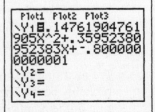

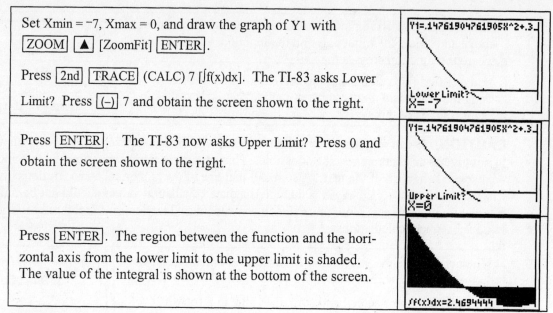

Set Xmin = ⁻7, Xmax = 0, and draw the graph of Y1 with ZOOM ▲ [ZoomFit] ENTER . Press 2nd TRACE (CALC) 7 [∫f(x)dx]. The TI-83 asks Lower Limit? Press (−) 7 and obtain the screen shown to the right.	Y1=.14761904761905X^2+.3 Lower Limit? X=⁻7
Press ENTER . The TI-83 now asks Upper Limit? Press 0 and obtain the screen shown to the right.	Y1=.14761904761905X^2+.3 Upper Limit? X=0
Press ENTER . The region between the function and the horizontal axis from the lower limit to the upper limit is shaded. The value of the integral is shown at the bottom of the screen.	∫f(x)dx=2.4694444

CAUTION: If you type in a value for the lower and/or upper limit (that is, the input at the endpoint) that is not visible on the graphics screen, you will get an error message when you attempt to evaluate the integral. Be certain that these limits are included in the interval from Xmin to Xmax before using this method.

6.5.2 FINDING THE AREA BETWEEN TWO CURVES The process of finding the area of the region between two functions uses many of the techniques presented in preceding sections. If the two functions intersect, you need to first find the input values of the point(s) of intersection. We illustrate these ideas as they are presented in Example 5 of Section 6.5 of the text.

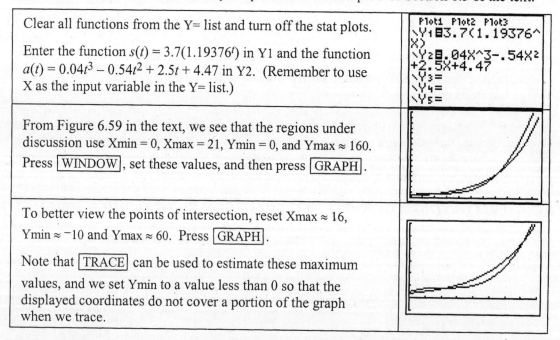

Clear all functions from the Y= list and turn off the stat plots. Enter the function $s(t) = 3.7(1.19376^t)$ in Y1 and the function $a(t) = 0.04t^3 - 0.54t^2 + 2.5t + 4.47$ in Y2. (Remember to use X as the input variable in the Y= list.)	Plot1 Plot2 Plot3 \Y1☐3.7(1.19376^X) \Y2☐.04X^3-.54X² +2.5X+4.47 \Y3= \Y4= \Y5=
From Figure 6.59 in the text, we see that the regions under discussion use Xmin = 0, Xmax = 21, Ymin = 0, and Ymax ≈ 160. Press WINDOW , set these values, and then press GRAPH .	
To better view the points of intersection, reset Xmax ≈ 16, Ymin ≈ ⁻10 and Ymax ≈ 60. Press GRAPH . Note that TRACE can be used to estimate these maximum values, and we set Ymin to a value less than 0 so that the displayed coordinates do not cover a portion of the graph when we trace.	

We next find the inputs of the points of intersection of the two functions. (These values will probably be the limits on the integrals we use to find the areas.) The method we use to find the first of the points is the intersection method that was discussed in Section 1.1.1i of this *Guide*.

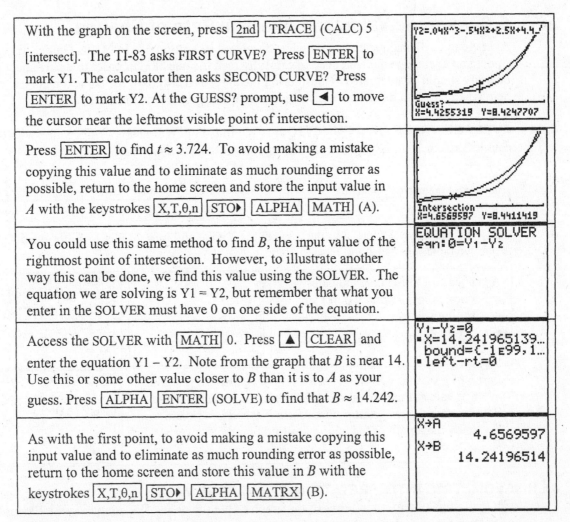

With the graph on the screen, press 2nd TRACE (CALC) 5 [intersect]. The TI-83 asks FIRST CURVE? Press ENTER to mark Y1. The calculator then asks SECOND CURVE? Press ENTER to mark Y2. At the GUESS? prompt, use ◄ to move the cursor near the leftmost visible point of intersection.	Y2=.04X^3-.54X2+2.5X+4.4_/ Guess? X=4.4255319 Y=8.4247707
Press ENTER to find $t \approx 3.724$. To avoid making a mistake copying this value and to eliminate as much rounding error as possible, return to the home screen and store the input value in A with the keystrokes X,T,θ,n STO► ALPHA MATH (A).	Intersection X=4.6569597 Y=8.4411419
You could use this same method to find B, the input value of the rightmost point of intersection. However, to illustrate another way this can be done, we find this value using the SOLVER. The equation we are solving is Y1 = Y2, but remember that what you enter in the SOLVER must have 0 on one side of the equation.	EQUATION SOLVER eqn:0=Y1-Y2
Access the SOLVER with MATH 0. Press ▲ CLEAR and enter the equation Y1 − Y2. Note from the graph that B is near 14. Use this or some other value closer to B than it is to A as your guess. Press ALPHA ENTER (SOLVE) to find that $B \approx 14.242$.	Y1-Y2=0 ▪X=14.241965139… bound={-1E99,1… ▪left-rt=0
As with the first point, to avoid making a mistake copying this input value and to eliminate as much rounding error as possible, return to the home screen and store this value in B with the keystrokes X,T,θ,n STO► ALPHA MATRX (B).	X→A 4.6569597 X→B 14.24196514

NOTE: Even though we have found the only intersection points that are useful in the context of this example, the SOLVER can easily find if there are any other intersection points for these two curves. Enter several different guesses, some smaller than A and some larger than B. (We can see from the graph that there are no other points of intersection between A and B.) You should find that these curves also intersect at $t \approx {-0.372}$ and $t \approx 28.077$.

Calculate the areas of the two regions enclosed by s and a (that is, regions R_1 and R_2 that are shown in Figure 6.60 in the text) as indicated to the right. Note that each answer should be positive because we are finding areas. If you obtain a negative answer, you probably entered the functions in the wrong order.	fnInt(Y1-Y2,X,10 ,B) 18.53830546 fnInt(Y2-Y1,X,B, 20) 79.40758948

If you are in doubt as to which function to enter first in the integral, you can press GRAPH and trace the graphs to see which function is on top of the other. Press ▲ and ▼ to have the cursor jump from one curve to the other and notice the equation at the top of the screen.

NOTE: The value $18.53830546 - 79.40758948 \approx {-60.869}$ could also have been calculated by evaluating fnInt(Y1 − Y2, X, 10, 20). Because the graph in Figure 6.60 shows that the area of R_2 is greater than the area of R_1, we see that fnInt(Y1 − Y2, X, 10, 20) should be negative.

6.6 Average Value and Average Rate of Change

Average rates of change are computed as discussed in Section 3.1.1a of this *Guide*. When finding an average value, you need to carefully read the question in order to determine which quantity should be integrated. Considering the units of measure in the context can be a great help when trying to determine which function to integrate to find an average value.

6.6.1a AVERAGE VALUE OF A FUNCTION
We illustrate finding an average value with the data in Table 6.21 in Example 1 of Section 6.6 of *Calculus Concepts*:

Time (number of hours after midnight)	7	8	9	10	11	12	13	14	15	16	17	18	19
Temperature (°F)	49	54	58	66	72	76	79	80	80	78	74	69	62

Clear any old data. Delete any functions in the Y= list and turn on Plot 1. Enter the data in the above table in lists L1 and L2.

A scatter plot of the data indicates an inflection point (around 9 p.m.) and no limiting values. Fit a cubic function to the data and paste it in Y1.		Plot1 Plot2 Plot3 \Y1■-.0352564102 564X^3+.71815684 315655X^2+1.5843 32334336X+13.689 310689299 \Y2= \Y3=

Part *b* of Example 1 asks for the average *temperature* (*i.e.*, the average value of the temperature) between 9 a.m. and 6 p.m. So, integrate the *temperature* between $x = 9$ and $x = 18$ and divide by the length of the interval to find that the answer is about 74.4°F.		fnInt(Y1,X,9,18)
		    669.8450612
		Ans/(18-9)
		    74.42722902

6.6.1b GEOMETRIC INTERPRETATION OF AVERAGE VALUE
What does the average value of a function mean in terms of the graph of the function? We continue with Example 1 of Section 6.6 of *Calculus Concepts* by considering the function and average value found in Section 6.6.1a. Have the unrounded cubic temperature function in Y1.

Enter 74.42722902 in Y2. You can either leave Ymin at the setting from the scatter plot or change Ymin to 0. Press GRAPH to see the graph of the average value and the function.		

We illustrate for only the second graph shown above, but what follows is true for both. The area of the rectangle whose height is the average temperature is $(74.42722902 - 0)(18 - 9) \approx$ 669.845. The area of the region between the temperature function and the input axis between 9 a.m. and 6 p.m. is this same value.		74.42722902*(18-9)
		    669.8450612
		fnInt(Y1,X,9,18)
		
		    669.8450612

To find the answer to part *d* of Example 1, enter and evaluate the expression shown to the right. The average rate of change of the temperature from 9 a.m. to 6 p.m. is about 0.98°F/hr.		Y1(18)-Y1(9)
		    8.857642358
		Ans/(18-9)
		    .9841824842

Chapter 7   Analyzing Accumulated Change: Integrals in Action

 ## 7.2 Streams in Business and Biology

You will find your calculator very helpful when dealing with streams that are accumulated over finite intervals. Finding either the future or present value of a continuous income stream is simply finding the value of a definite integral. However, the technique used with discrete income streams involves the TI-83's sequence function that was introduced in Section 2.1.1a. The details are presented in Section 7.2.2 of this *Guide*.

7.2.1 DETERMINING THE FLOW-RATE FUNCTION FOR AN INCOME STREAM   The TI-83 can often help you to find the equation for an income stream flow rate. Note that it is *not* necessary to use the calculator to find such an equation – we present this as a technique to use only if you find writing the equation from the word description difficult. We illustrate these ideas as they are given in Example 1 of Section 7.2 of *Calculus Concepts*.

In Example 1, part *a*, we are told that the business's profit remains constant. Clear lists L1 and L2. In L1 enter two possible input values for the time involved. (You might use different years than the ones shown here.) In L2 enter the amount invested: 10% of the constant profit.	L1 L2 L3 1 1 57900 ------ 2 57900 ------ L1(3)=
You need to remember that a *constant output* means a *linear flow rate*. Fit a linear function to these two data points to find that $R(t) = 57{,}000$ dollars per year.	LinReg y=ax+b a=0 b=57900

CAUTION: If you attempt to draw a scatter plot on the calculator, you will get an error message because the TI-83, using the output data in L2, sets Ymin = Ymax. (You need to draw the scatter plot using paper and pencil.) If you want to see the horizontal line graph on the TI-83, change Ymin and Ymax so that 57,900 is between the two values and press ⎡GRAPH⎤.

In Example 1, part *b*, we are told that the business's profit grows by $50,000 each year. The first year's profit (which determines the initial investment at $t = 0$) is $579,000. Reason that if the profit grows by $50,000 each year, the next year's profit will be $579,000 + 50,000 = $629,000. Enter these values in L1 and L2.	L1 L2 L3 2 0 579000 ------ 1 629000 ------ L2(3) =
You need to remember that *constant growth* means a *linear flow rate*. Fit a linear function to these two data points. Next, carefully read the problem once more. Note that only 10% of the profit is invested. Thus, the linear flow rate function is $R(t) = 0.10(50{,}000t + 579{,}000)$ dollars per year t years after the first year of business.	LinReg y=ax+b a=50000 b=579000

| In Example 1, part *c*, we are told that the business's profit grows by 17% each year. The first year's profit (which determines the initial investment at *t* = 0) is $579,000. Reason that if the profit grows by 17% each year, the next year's profit will be $579,000 + 0.17(579,000) = $677,430. Enter these values in L1 and L2. | ```
L1      L2      L3    2
0       579000  ------
1       677430
------
L2(3) =
``` |
| You need to remember that a *constant percentage growth* means an *exponential flow rate*. Fit an exponential function to these two data points. Now, carefully read the problem once more. Note that only 10% of the profit is invested. Thus, the exponential flow rate function is $R(t) = 0.10(579,000)(1.17^t)$ dollars per year *t* years after the first year of business. | ```
ExpReg
y=a*b^x
a=579000
b=1.17
``` |

Part *d* of Example 1 gives data that describe the growth of the business's profit. Refer to the material in Section 2.2.1 of this *Guide* to review how to fit a log function to these data points.

NOTE: If you forget which type of growth gives which function, simply use what you are told in the problem and fill in the lists with approximately five data points. Draw a scatter plot of the data and it should be obvious from the shape which function to fit to the data.

| For instance, return to Example 1, part *b*, in which we are told that the business's profit grows by $50,000 each year. Note that we assume throughout this section that initial investments are made at time *t* = 0. Simply add $50,000 to each of the previous year's profit to obtain about five data points. | ```
L1      L2      L3    2
0       579000  ------
1       629000
2       679000
3       729000
4       779000
------   ------
L2(5) =...000+50000
``` |
| Draw a scatter plot, and observe that the points fall in a line! Note that if you run program DIFF, the first differences are constant at 50,000.

Next, find a linear equation for the data and find that the flow rate function is $R(t) = 0.10(50,000t + 579,000)$ dollars per year *t* years after the first year of business. | |

7.2.2 FUTURE VALUE OF A DISCRETE INCOME STREAM

The future value of a discrete income stream is found by adding, as *d* increases, the terms of $R(d)\left(1 + \frac{r}{n}\right)^{D-d}$ where $R(d)$ is the value per period of the *d*th deposit, 100*r*% is the annual percentage rate at which interest is earned when the interest is compounded once in each deposit period, *n* is the number of times interest is compounded (and deposits are made) during the year, and *D* is the total number of deposit periods. It is assumed that initial deposits are made at time *t* = 0 unless it is otherwise stated. We use the situation in Example 4 in Section 7.2 of *Calculus Concepts*:

> When you graduate from college (say, in 3 years), you would like to purchase a car. You have a job and can put $75 into savings each month for this purchase. You choose a money market account that offers an APR of 6.2% compounded quarterly.

Part *a* asks how much will you have deposited in 3 years. No interest is involved in this calculation. There are 12 months in each year and $75 is deposited each month for 3 years.

A total of $2700 is deposited. | ```
12*3*75
              2700
``` |

Part b of Example 4 asks for the future value of the deposits at the end of 3 years. Because the APR is 6.2% compounded quarterly, $r = 0.062$, $n = 4$, and $D = 3 \text{ years} \cdot 4 \frac{\text{deposits}}{\text{year}} = 12$ deposits. A constant \$75 each month is deposited, so $R(d) = \$75(3) = \$225$ deposited each quarter. The 3-year future value is given by $\sum_{d=0}^{11} 225\left(1 + \frac{0.062}{4}\right)^{12-d}$.

The TI-83 sequence command can be used to find this sum. The syntax for this is

seq(formula, variable, first value, last value, increment)

When applied to discrete income streams, the *formula* for the sequence is $R(d)\left(1 + \frac{r}{n}\right)^{D-d}$.

For convenience, we enter the formula in Y1. You must enclose the exponent in parentheses. Otherwise, the TI-83 will assume that only 12 is in the exponent.	Plot1 Plot2 Plot3 \Y1▪225(1+.062/4)^(12-X) \Y2= \Y3= \Y4= \Y5= \Y6=
Return to the home screen and enter seq(Y1, X, 0, 11, 1). Recall that the sequence command is accessed by pressing [2nd] [STAT] (LIST) [▶] [OPS] 5 [seq(]. If you want to see the 12 future values in this list, scroll through the list with [▶].	seq(Y1,X,0,11,1) {270.6086352 26…

NOTE: When you use the sequence command for discrete income streams, the first value is always 0. The last value is always $D - 1$ because we start counting at 0, not 1. The increment will always be 1 because of the way the formula is designed.

What are the values in the list that result when you use the sequence command? The first value (approximately \$270.61) is the 3-year future value of the first 3 months deposits (\$225). The second value (approximately \$266.48) is the 3-year future value of the second 3 months of deposits (\$225), and so forth.

To find the future value of all the deposits, simply sum the values in the list. To do this, use the sum command to add the values in the list (the previous ANS). The sum command is accessed with [2nd] [STAT] (LIST) [▶] [▶] [MATH] 5 [sum(]. Store this value in F for use in the next part of the problem.	seq(Y1,X,0,11,1)
	{270.6086352 26…
	sum(Ans)
	    2988.101231
	Ans→F
	    2988.101231

Part c of Example 4 asks for the present value of the amount you would have to deposit now to achieve the same future value as was found above. To find the present value P, solve the equation $P\left(1 + \frac{r}{n}\right)^D = \text{Future value}$. Enter $P\left(1 + \frac{0.062}{4}\right)^{12} - F = 0$ as the equation in the SOLVER.	EQUATION SOLVER eqn:0=P(1+.062/4)^12-F
Press [ENTER]. Enter a guess for P, and do not change the value of F. (There is only one solution, so any positive number should work as a guess.) With the cursor on the P line, press [ALPHA] [ENTER] (SOLVE) to find $P \approx \$2484.48$.	P(1+.062/4)^1…=0 ▪P=2484.4838243… F=2988.1012310… bound={-1ᴇ99,1… ▪left-rt=0

| Part d of Example 4 asks for the future value if the monthly interest rate is 0.5% instead of the previous APR compounded quarterly. Press $\boxed{Y=}$ and make the appropriate changes in the function in Y1. Don't forget that $R(d)$ is the amount deposited per interest compounding period. | ```
Plot1 Plot2 Plot3
\Y1■75(1+.005)^(
36-X)
\Y2=
\Y3=
\Y4=
\Y5=
\Y6=
``` |
| Press $\boxed{2nd}$ $\boxed{ENTER}$ (ENTRY) until the sequence command reappears on the home screen. Change *last value* to 35 and press $\boxed{ENTER}$. Sum the resulting sequence to find that the 3-year future value is about $2964.96. | ```
seq(Y1,X,0,35,1)
(89.75103936 89…
sum(Ans)
     2964.958912
``` |

NOTE: You can use any letter for the input variable in the formula if you enter the function on the home screen, but if you are entering the formula in the Y= list, you must use X.

EXTENSION: The TI-83 calculator has a built-in financial menu accessed with $\boxed{2nd}$ $\boxed{x^{-1}}$ (FINANCE). The items in this menu can be extremely useful in other courses with discrete financial applications. Consult your *Owners' Manual* for more details and instructions.

7.3 Integrals in Economics

Consumers' surplus and producers' surplus (when they are defined by definite integrals) are easy to find using fnInt. You should draw graphs of the demand and supply functions and think of the economic quantities in terms of area so to understand the questions being asked.

7.3.1 CONSUMER ECONOMICS   We illustrate how to find the consumers' surplus and other economic quantities when the demand function intersects the input axis as given in Example 1 of Section 7.3 of *Calculus Concepts*:

> Suppose the demand for a certain model of minivan in the United States can be described as $D(p) = 14.12(0.933^p) - 0.25$ million minivans when the market price is p thousand dollars per minivan.

We first draw a graph of the demand function. This is not asked for, but it will really help your understanding of the problem. Read the problem to see if there are any clues as to how to set the horizontal view for the graph. The price cannot be negative, so $p \geq 0$. There is no price given in the remainder of the problem, so just guess a value with which to begin.

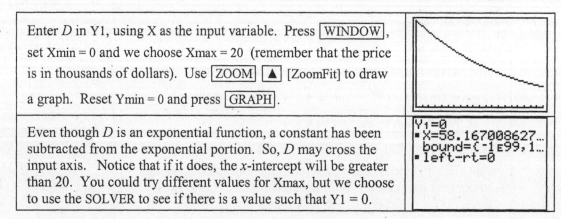

| Enter D in Y1, using X as the input variable. Press $\boxed{WINDOW}$, set Xmin = 0 and we choose Xmax = 20 (remember that the price is in thousands of dollars). Use $\boxed{ZOOM}$ $\boxed{\blacktriangle}$ [ZoomFit] to draw a graph. Reset Ymin = 0 and press $\boxed{GRAPH}$. | |
| Even though D is an exponential function, a constant has been subtracted from the exponential portion. So, D may cross the input axis. Notice that if it does, the x-intercept will be greater than 20. You could try different values for Xmax, but we choose to use the SOLVER to see if there is a value such that Y1 = 0. | ```
Y1=0
■X=58.167008627…
 bound={-1E99,1…
■left-rt=0
``` |

Customers will not purchase this model minivan if the price per minivan is more than about \$58.2 thousand. Store this value in P for later use and set Xmax = P. Redraw the graph with GRAPH. (Be sure to label the axes with variables and units of measure when you copy this graph to your paper.)

Note that the answer to Example 1, part c, is $p \approx \$58.2$ thousand.

Part a of Example 1 asks at what price consumers will purchase 2.5 million minivans. Look at the labels on your graph and note that 2.5 is a value of D, not p. You therefore need to find the price (an input). Return to the solver and edit the equation so to solve Y1 − 2.5 = 0. You can trace the graph for a guess, but there is only one answer, so any reasonable guess will suffice.

Part b asks for the consumers' expenditure when purchasing 2.5 million minivans. First, store this market price in M for future use. (Also label this value M on your hand-drawn graph.) The consumers' expenditure is $price * quantity$ = area of the rectangle with height = 2.5 million minivans and width $\approx \$23.59$ thousand per minivan. The area is about \$59 billion.

The consumer's surplus in part d of Example 1 is the area under the demand curve to the right of M ($M \approx 23.5903$) and to the left of P ($P \approx 58.1670$). The surplus is about \$27.4 billion.

The TI-83 draws the consumers' surplus if you use the methods illustrated in Section 6.5.1b.

First, reset Ymin to −2 to have more room at the bottom of the graph. Draw the graph of D. Then, with the graph on the screen, press 2nd TRACE (CALC) 7 [∫f(x)dx]. At the LOWER LIMIT? prompt, press ALPHA ÷ (M) ENTER. At the UPPER LIMIT? prompt, press ALPHA 8 (P) ENTER. You should then see the screen shown to the right.

CAUTION: Unless the values you enter for the upper and lower limits are visible on the graphics screen, you will get an error message using the above instructions. If you do as instructed and use P, make sure you used P or a value larger than P when setting Xmax.

7.3.2 **PRODUCER ECONOMICS AND SOCIAL GAIN**   We illustrate how to find producers' surplus and other economic quantities as indicated in Example 3 of Section 7.3 in *Calculus Concepts*:

> The demand and supply functions for the gasoline example in the text are given by $D(p) = 5.43(0.607^p)$ million gallons and $S(p) = 0$ million gallons for $p < 1$ and $S(p) = 0.792p^2 − 0.433p + 0.314$ million gallons for $p \geq 1$ when the market price of gas is p dollars per gallon.

First, let's find the market equilibrium point. This point can be found graphically using the intersection command and the graphs of the functions, but we choose to use the SOLVER.

Enter D in Y1. Enter S (for $p \geq 1$) in Y2 as shown to the right. Use X as the input variable in each function. Recall that you need to enclose the piecewise function and its input in parentheses. Access the inequality symbol with 2nd MATH 4 [≥]. Because we intend to draw a graph of S, put Y2 in Dot mode.	Plot1 Plot2 Plot3 \Y1■5.43(.607^X) \Y2■(.792X²−.433 X+.314)(X≥1) \Y3= \Y4= \Y5=
Access the SOLVER and enter the equation Y1 − Y2 = 0. Enter a guess for the equilibrium point. With the cursor on the X line, press ALPHA ENTER (SOLVE) to find X ≈ 1.8331. Store this value in some memory location, say E.	Y1−Y2=0 •X=1.8311107618… bound={−1ᴇ99, 1… •left−rt=0

NOTE: Is X ≈ 1.8331 the only solution to the equation Y1 − Y2 = 0? Enter more guesses for X, some large and some small. Each time the solution shown above results, so it appears that there is only one solution.

We next draw graphs of the demand and supply functions. How do we set the window to draw the graphs? Notice that because D is an exponential function, it will never cross the input axis. Because S is a concave-up parabola, it also increases when X > 1. Therefore, just try several different Xmax values and choose an Xmax so that the graphs can be viewed.

Press WINDOW, set Xmin = 0 and we choose Xmax = 3. Use ZOOM ▲ [ZoomFit] to draw the graphs of D and S. If Ymin is not 0, set it to 0 and press GRAPH.	

Note that there is only one point of intersection for the two functions, the market equilibrium point with input X ≈ 1.8331. Recall that we stored the complete value of X in E.

Total social gain = producers' surplus + consumers' surplus at the equilibrium price. Shade this area on your graph and then use the graph to write the integrals that give this area:

$$\text{Social gain} = \int_1^E S(p)\,dp + \int_E^\infty D(p)\,dp \text{ where } E \approx 1.8331$$

Find the producer's surplus as shown to the right.
 	
The TI-83 does not calculate the value of improper integrals. However, as a check on your algebraic work to determine the value of the consumer's surplus, you can do the following.
	X→E
	        1.831110762
	fnInt(Y2,X,1,E)
	        1.108419968

Enter fnInt(Y1, X, E, X) in a location of the Y= list, say Y3. Turn off Y1 and Y2. Set the TI-83 TABLE to ASK and enter increasingly larger values of X such as those shown to the right.	Plot1 Plot2 Plot3 \Y1=5.43(.607^X) \Y2=(.792X²−.433 X+.314)(X≥1) \Y3■fnInt(Y1,X,E ,X) \Y4=	X | Y3 5 | 3.4638 50 | 4.3601 180 | 4.3601 395 | 4.3601 800 | 4.3601 Y3=4.3601023937

It appears that $\lim\limits_{N \to \infty} \int_E^N D(p)\,dp \approx 4.360$, giving social gain ≈ 1.108 + 4.360 ≈ $5.5 million.

NOTE: The remainder of the material in this *Guide* refers to the complete text for *Calculus Concepts* (that also contains Section 7.4 and Chapters 8-11.)

 ## 7.4 Probability Distributions and Density Functions

Most of the applications of probability distributions and density functions use technology techniques that have already been discussed. Probabilities are areas whose values can be found by integrating the appropriate density function. A cumulative density function is an accumulation function of a probability density function.

Your calculator's numerical integrator is especially useful for finding means and standard deviations of some probability distributions because those integrals often involve expressions for which we have not developed algebraic techniques for finding antiderivatives. The TI-83 calculator contains many built-in statistical menus that deal with probability distributions and density functions. These are especially useful when you take a statistics course, but some can be used with Section 7.4 in this course. Consult your TI-83 *Owners' Manual* for more details and instructions for use of these features.

7.4.1a NORMAL PROBABILITIES   The normal density function is the most well known and widely used probability distribution. If you are told that a random variable x has a normal distribution $N(x)$ with mean μ and standard deviation σ, the probability that x is between two real numbers a and b is given by

$$\int_a^b N(x)\,dx = \int_a^b \frac{1}{\sigma\sqrt{2\pi}}\, e^{\frac{-(x-\mu)^2}{2\sigma^2}}\, dx$$

We illustrate these ideas with the situation in Example 8 of Section 7.4 of *Calculus Concepts*. In that example, we are given that the distribution of the life of the bulbs, with the life span measured in hundreds of hours, is modeled by a normal density function with $\mu = 900$ hours and $\sigma = 100$ hours. Carefully use parentheses when entering the normal density function.

The probability that a light bulb lasts between 9 and 10 hundred hours is $$P(9 \le x \le 10) = \int_9^{10} \frac{1}{\sqrt{2\pi}}\, e^{-(x-9)^2/2}\, dx$$ The value of this integral must be a number between 0 and 1.	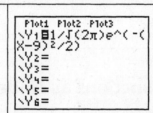

7.4.1b VIEWING NORMAL PROBABILITIES   We could have found the probability that the light bulb life is between 900 and 1000 hours graphically by using the CALC menu.

To draw a graph of the normal density function, note that nearly all of the area between this function and the horizontal axis lies within three standard deviations of the mean. Set Xmin = 9 – 3 = 6 and Xmax = 9 + 3 = 12. Use ZoomFit to draw the graph. Reset Ymin to ⁻0.1 in preparation for the next step, and press GRAPH.	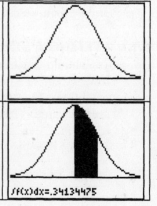
With the graph on the screen, press 2nd TRACE (CALC) 7 [∫f(x)dx]. At the Lower Limit? prompt, enter 9. At the Upper Limit? prompt, enter 10. The probability is displayed as an area and the value is printed at the bottom of the graphics screen.	

Chapter 8   Repetitive Change: Cycles and Trigonometry

 8.1 Functions of Angles: Sine and Cosine

Before you begin this chapter, go back to the first page of the *Graphing Calculator Instruction Guide* and check the basic setup, the statistical setup, and the window setup. If these are not set as specified in Figures 1, 2, and 3, you will have trouble using your calculator in this chapter. Pay careful attention to the third line in the MODE screen in the basic setup. The Radian/Degree mode setting affects the TI-83's interpretation of the ANGLE menu choices. The calculator's MODE menu should always be set to Radian unless otherwise specified. (Note that calculator instructions for material that is in *Appendix A: Trigonometry Basics* are on the *Calculus Concepts* CD-ROM and web site.)

8.1.1 FINDING OUTPUTS OF TRIG FUNCTIONS WITH RADIAN INPUTS   It is essential that you have the correct mode set when evaluating trigonometric function outputs. The angle setting in the MODE menu must be Radian for all applications in Chapter 8. We illustrate how to evaluate trig functions with the following example.

Find $\sin\frac{9\pi}{8}$ and $\cos\frac{9\pi}{8}$. Because these angles are in radians, be certain that Radian is chosen in the third line of the MODE screen. The $\boxed{\text{SIN}}$ and $\boxed{\text{COS}}$ keys are above the $\boxed{,}$ and $\boxed{(}$ keys on the TI-83 keyboard.	`sin(9π/8)`
	`        -.3826834324`
	`cos(9π/8)`
	`        -.9238795325`

It is also essential that you use parentheses to indicate the order of operations. When you press any of the trig function keys, the left parenthesis automatically appears and cannot be deleted. If anything follows the angle, the right parenthesis is necessary to show the end of the input of the trig function.	`sin(9π/8`
	`        -.3826834324`
	`sin(9π/8)+3`
	`        2.617316568`
	`sin(9π/8+3`
	`        .2484758395`

 8.2 Cyclic Functions as Models

We now introduce another model – the sine model. As you might expect, this function should be fit to data that repeatedly varies between alternate extremes. The form of the sine model is given by $f(x) = a \sin(bx + h) + k$ where $|a|$ is the amplitude, b is the frequency (where $b > 0$), $2\pi/b$ is the period, $|h|/b$ is the horizontal shift (to the right if $h < 0$ and to the left if $h > 0$), and k is the vertical shift (up if $k > 0$ and down if $k < 0$). *Note:* The TI-83 uses the c when we use h and d when we use k.

8.2.1 FITTING A SINE MODEL TO DATA   Before fitting any model to data, remember that you should construct a scatter plot of the data and observe what pattern the data appear to follow. Example 2 in Section 8.2 asks you to find a sine model for cyclic data with the hours of daylight on the Arctic Circle as a function of the day of the year on which the hours of daylight are measured. (January 1 is day 1.) These data appear in Table 8.3 of *Calculus Concepts*.

Day of the year	-10	81.5	173	264	355	446.5	538	629	720	811.5
Hours of daylight	0	12	24	12	0	12	24	12	0	12

Clear any old data. Delete any functions in the Y= list and turn on Plot 1. Enter the data in the above table in lists L1 and L2. Construct a scatter plot of the data. When using the sine regression in the TI-83, it is sometimes necessary to have an estimate of the period of the data.

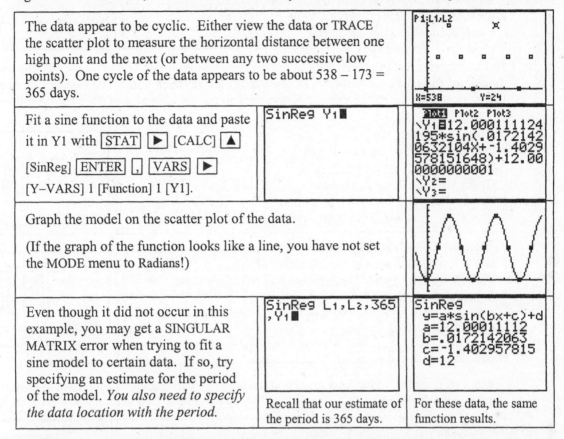

The data appear to be cyclic. Either view the data or TRACE the scatter plot to measure the horizontal distance between one high point and the next (or between any two successive low points). One cycle of the data appears to be about 538 − 173 = 365 days.		
Fit a sine function to the data and paste it in Y1 with $\boxed{\text{STAT}}$ $\boxed{\blacktriangleright}$ [CALC] $\boxed{\blacktriangle}$ [SinReg] $\boxed{\text{ENTER}}$ $\boxed{,}$ $\boxed{\text{VARS}}$ $\boxed{\blacktriangleright}$ [Y−VARS] 1 [Function] 1 [Y1].		
Graph the model on the scatter plot of the data. (If the graph of the function looks like a line, you have not set the MODE menu to Radians!)		
Even though it did not occur in this example, you may get a SINGULAR MATRIX error when trying to fit a sine model to certain data. If so, try specifying an estimate for the period of the model. *You also need to specify the data location with the period.*	Recall that our estimate of the period is 365 days.	For these data, the same function results.

NOTE: If you do not think the original function the calculator finds fits the data very well, try specifying a period and see if a better-fitting equation results. It didn't here, but it might with a different set of data. If you still cannot find an equation, you can tell the TI-83 how many times to go through the routine that finds the equation. This number of iterations is 3 if not specified. The number should be typed before L1 when initially finding the equation.

8.3 Rates of Change and Derivatives

All the previous techniques given for other functions also hold for the sine model. You can find intersections, maxima, minima, inflection points, derivatives, integrals, and so forth.

8.3.1 DERIVATIVES OF SINE AND COSINE FUNCTIONS
Evaluate nDeriv at a particular input to find the value of the derivative of the sine function at that input. We illustrate with Example 1 in Section 8.3 of *Calculus Concepts*:

The calls for service made to a county sheriff's department in a certain rural/suburban county can be modeled as $c(t) = 2.8 \sin(0.262t + 2.5) + 5.38$ calls during the tth hour after midnight.

CAUTION: Because the TI-83 automatically inserts a left parentheses when you press the sine and cosine keys, be very careful that you do not type an extra parentheses when entering a trig function in the graphing list. If you do, you will change the order of operations and not have the correct equation entered. The function (see below) that is entered in Y2 is <u>incorrect</u>.

The correct way to enter the function c is shown in Y1. When these two equations are graphed, you can see that they are entirely different functions! The correct function is the darker one.	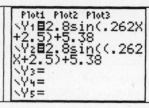	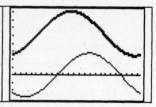

Delete the incorrect function in Y2. We now return to Example 1. Part *a* asks for the average number of calls the county sheriff's department receives each hour. The easiest way to obtain this answer is to remember that the parameter k in the sine function is the average value.

| You can also find the average value over one period of the function using the methods of Section 6.6.1a of the *Graphing Calculator Instruction Guide*. Enter the function $c$ in Y1 and type in the quotient shown to the right. | Average Value = $$\dfrac{\displaystyle\int_0^{2\pi/0.262} c(t)\,dt}{2\pi/0.262 - 0}$$ | ```
fnInt(Y₁,X,0,2π/
.262)
        129.0211334
Ans/(2π/.262)
               5.38
``` |
|---|---|---|
| Enter the TI-83's numerical derivative in Y2. Enter your answer to part *b*, the formula for $c'(t)$, in Y3. As discussed in Sections 4.3.2b/4.3.2c of the *Graphing Calculator Instruction Guide*, use the TABLE or the graphs of Y2 and Y3 to check your answer. | ```
Plot1 Plot2 Plot3
\Y₁=2.8sin(.262X
+2.5)+5.38
\Y₂∎nDeriv(Y₁,X,
X)
\Y₃∎.7336cos(.26
2X+2.5)
\Y₄=
``` | | X | Y₂ | Y₃ |
|---|---|---|
0	-.5877	-.5877
2	-.7285	-.7285
5	-.5757	-.5757
18	.43691	.43691
22	-.2924	-.2924
34	.29405	.29405

X= |

Part *c* of Example 1 asks how quickly the number of calls received each hour is changing at noon and at midnight.

| To answer these questions, simply evaluate Y2 (or your derivative in Y3) at 12 for noon and 0 (or 24) for midnight. (You were not told if "midnight" refers to the initial time or 24 hours after that initial time.) | ```
Y₂(12)
         .588774165
Y₂(0)
        -.5877189497
Y₂(24)
        -.5898259682
``` |
|---|---|

 ## 8.5 Accumulation in Cycles

As with the other functions we have studied, applications of accumulated change with the sine and cosine functions involve the calculator's numerical integrator fnInt.

8.5.1 INTEGRALS OF SINE AND COSINE FUNCTIONS
We illustrate the process of determining accumulated change with Example 1 in Section 8.5 of *Calculus Concepts*.

| Enter the rate of change of temperature in Philadelphia on August 27, 1993 in Y1. Find the accumulated change in the temperature between 9 a.m. and 3 p.m. using fnInt. | ```
Plot1 Plot2 Plot3
\Y₁∎2.733cos(.28
5X-2.93)
\Y₂=
\Y₃=
\Y₄=
\Y₅=
\Y₆=
``` | ```
fnInt(Y₁,X,9,15)
        12.76901162
``` |
|---|---|---|

The temperature increased by about 13°F between 9 a.m. and 3 p.m.
```

Chapter 9   Ingredients of Multivariable Change: Models, Graphs, Rates

 ## 9.1 Multivariable Functions and Contour Graphs

Because any program that we might use to graph[1] a three-dimensional function would be fairly involved and take a long time to execute, we do not graph three-dimensional functions. Instead, we discover information about three-dimensional graphs using their associated contour curves.

9.1.1 SKETCHING CONTOUR CURVES

When given a multivariable function with two input variables, you can draw contour graphs using the three-step process described below. We illustrate with a function that gives body heat loss due to the wind. This function, H, appears in Example 2 of Section 9.1 of *Calculus Concepts*:

$$H(v, t) = (10.45 + 10\sqrt{v} - v)(33 - t) \text{ kilogram-calories}$$

per square meter of body surface area per hour for wind speed v in meters per second.

Step 1: Set $H(v, t) = 2000$. Because we will use the SOLVER to find the value of t at various values of v, write the function as $H(v, t) - 2000 = 0$.

Step 2: Choose values for v and solve for t to obtain points on the 2000 constant-contour curve. Obtain guesses for the values of v and t from Table 9.2 in the text.

Enter the $H(v, t)$ formula in Y1. Be sure that you use the letters V and T, not X. (We are only using Y1 as a "holding place" for the formula.) Go to the SOLVER and enter the equation Y1 – 2000. Press ENTER.	```Plot1 Plot2 Plot3 \Y1◻(10.45+10√(V)-V)(33-T) \Y2= \Y3= \Y4= \Y5= \Y6=```	```EQUATION SOLVER eqn:0=Y1-2000```
Have the cursor on the V line and type 5. Press ▼ or ENTER to move the cursor to the T line and solve for T with ALPHA ENTER (SOLVE).	```Y1-2000=0 V=5▮ T=0 bound={-1E99, 1…```	```Y1-2000=0 V=5 •T=-38.91481891… bound={-1E99, 1… •left-rt=0```

WARNING: The cursor must be on the line corresponding to the unknown variable for the SOLVER to solve the equation for that variable. If you do not have a table of values for the quantities, you should enter several different guesses for the unknown variable to determine whether there is more than one solution to the equation.

Press ▲, enter another value for v, say $v = 10$ years. Move the cursor to the T line and press ALPHA ENTER (SOLVE). Enter $v = 15$ and repeat the procedure to solve for t.	```Y1-2000=0 V=10 •T=-29.35818073… bound={-1E99, 1… •left-rt=0```	```Y1-2000=0 V=15 •T=-25.51403583… bound={-1E99, 1… •left-rt=0```

---

[1] There are many TI-83 programs, including one that will graph a multivariable function with two input variables, available at the Texas Instruments web site with address http://education.ti.com.

| Repeat the procedure for $v = 20$ and $v = 25$. Make a table of the values of $v$ and $t$ as you find these values. | ```
Y1-2000=0
V=20
▪T=-23.86444952…
 bound={-1ᴇ99,1…
▪left-rt=0
``` | ```
Y1-2000=0
V=25
▪T=-23.41748942…
 bound={-1ᴇ99,1…
▪left-rt=0
``` |
|---|---|---|

Step 3: Plot the points obtained in Step 2 with pencil and paper. You need to find as many points as it takes to see the pattern the points are indicating when you plot them. Connect the points you have plotted with a smooth curve.

You must have a function given to draw a contour graph using the above method. Even though there may be several functions that seem to fit the data points obtained in Step 2, their use would be misleading because the real best-fit function can only be determined by substituting the appropriate values in a multivariable function. The focus of this section is to use contour graphs to study the relationships between input variables, not to find the equation of a function to fit a contour curve. Thus, we always sketch the contours on paper rather than with the TI-83.

 ## 9.2 Cross-Sectional Models and Rates of Change

For a multivariable function with two input variables, obtain a cross-sectional model by entering the data in lists L1 and L2 and then fitting the appropriate function as indicated in previous chapters of this *Guide*. Unless you are told otherwise, we assume that the data are given in a table with the values of the *first* input variable listed *horizontally* across the top of the table and the values of the *second* input variable listed *vertically* down the left side of the table.

9.2.1a FINDING A CROSS-SECTIONAL MODEL FROM DATA (HOLDING THE FIRST INPUT VARIABLE CONSTANT) We are to find the cross-sectional model $E(0.8, n)$ using the elevation data that appear in Table 9.4 of Section 9.2 in *Calculus Concepts*. However, the data you use in the activities will be obtained from a multivariable table. So we digress for a moment to understand how to find the data you need in such a table.

Refer to Table 9.3 on page 622 of the text. Remember that "rows" (containing the n data) go from left to right horizontally and "columns" (containing the e data) go from top to bottom vertically. In $E(0.8, n)$, e is constant at 0.8 and n varies. Thus, choose the values for n that appear on the left side of the table (vertically) in L1 and the elevations E in the $e = 0.8$ column of Table 9.3 in L2. *In general, the inputs that you enter in L1 are either across the top or down the left side of the table, and the outputs that you enter in L2 will always be in the main body of any multivariable data table in this text.* Thus the data we enter are these values:

n (miles)	1.5	1.4	1.3	1.2	1.1	1.0	0.9	0.8
Elevation (feet above sea level)	797.6	798.1	798.5	798.9	799.2	799.5	799.7	799.9
n (miles)	0.7	0.6	0.5	0.4	0.3	0.2	0.1	0.0
Elevation (feet above sea level)	800.0	800.1	800.1	800.1	800.0	799.9	799.7	799.5

After entering the data, clear any functions from the Y= list, and turn on Plot1. Draw a scatter plot of the data with ZOOM 9 [ZoomStat].	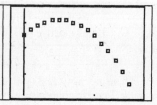

The data appear to be quadratic. Fit a quadratic function and copy it to the Y= list. (Refer to Section 2.4.1b of this *Guide*.) Overdraw the function on the scatter plot with GRAPH .

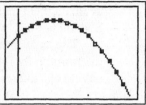

CAUTION: Because you will often be asked to find several different cross-sectional models using the same data table, calling different variables by the same names *x* and *y* would be very confusing. It is very important that you call the variables by the names that have been assigned in the problem. Remember that when finding or graphing a function, the TI-83 always calls the input variable X and the output variable Y. When working with multivariable functions, you must translate the calculator's equation $Y1 \approx -2.5X^2 + 2.497X + 799.490$ into the symbols that are used in the application. You should write the cross-sectional function as $E(0.8, n) = -2.5n^2 + 2.497n + 799.490$. Don't forget to completely describe (including units) all of the variables.

9.2.1b FINDING A CROSS-SECTIONAL MODEL FROM DATA (HOLDING THE SECOND INPUT VARIABLE CONSTANT)

The only difference in this model and the one in the previous section of this *Guide* is that the second input, instead of the first, is held constant. Refer again to Table 9.3 on page 622 of the text. Because we are now finding the cross-sectional model $E(e, 0.6)$, $n = 0.6$ and the inputs are the values of *e* that are across the top of the table. Enter these values in L1. (See the first box below for a shortcut.) The outputs are the elevations *E* obtained in the $n = 0.6$ mile row in Table 9.3. Enter these outputs in L2.

NOTE: You may find it helpful to place a piece of paper or a ruler under the row (or to the right of the column) in which the data appear to help avoid entering an incorrect value.

Because the input values for this function are the same as the input values in the last section of this *Guide* (but in reverse order) you can sort L1 in ascending order to avoid re-entering the data. Press 2nd STAT (LIST) ▶ [OPS] 1 [SortA(] 2nd 1 (L1) ENTER . Next, enter the output in L2.	SortA(L₁) Done
After entering the data, clear any functions from the Y= list, and turn on Plot1. Draw a scatter plot of the data with ZOOM 9 [ZoomStat].	(scatter plot and data table: L1 0,.1,.2,.3,.4,.5,.6 ; L2 802.8, 801.6, 800.8, 800.3, 800, 799.9, 799.9 ; L2(1)=802.8)
There is an inflection point and no evidence of limiting values, so the data appear to be cubic. Fit a cubic function and copy it to the Y= list. (Refer to Section 2.4.2 of this *Guide*.) Draw the cubic function on the scatter plot with GRAPH .	CubicReg y=ax³+bx²+cx+d a=-10.12410362 b=21.34716032 c=-13.97232949 d=802.8093911

9.2.2 EVALUATING OUTPUTS OF MULTIVARIABLE FUNCTIONS

As is the case with single-variable functions, outputs of multivariable functions are found by evaluating the function at the given values of the input variables. The main difference is that you usually will not be using X as the input variable symbol. One way to find multivariable function outputs is to evaluate them on the home screen. We illustrate with the investment function in Example 1 of Section 9.2 in *Calculus Concepts*.

The answer to part *a* of Example 1, as derived from the compound interest formula, uses the formula for the accumulated amount of an investment of *P* dollars for *t* years in an account paying 6% interest compounded quarterly:

$$A(P, t) = P\left(1 + \frac{0.06}{4}\right)^{4t} \text{ dollars}$$

When 10 is substituted for *t*, the cross-sectional function becomes $A(P, 10) \approx 1.814018409P$. Part *b* of Example 1 asks for $A(5300, 10)$. Even though it is simplest here to substitute 5300 for *P* in $A(P, 10) \approx 1.814018409P$, we return to the original function to illustrate evaluating multivariable formulas on the calculator.

To find the output on the home screen, type the formula for $A(P, t)$, substituting $P = 5300$ and $t = 10$. Press ENTER.	`5300(1+.06/4)^(4` `*10)` `         9614.297566`
Again, be warned that you must carefully use the correct placement of parentheses.	

Even though it is not necessary in this example, you may encounter activities in this section in which you need to evaluate a multivariable function at several different inputs. You could use what is shown above, but there are easier methods than individually entering each calculation. You will also use the techniques shown below in later sections of this chapter. When evaluating a multivariable function at several different input values, you may find it more convenient to enter the multivariable function in the graphing list.

Clear any previously-entered equations. Enter in Y1 the function $A(P, t) = P\left(1 + \frac{0.06}{4}\right)^{4t}$ with the keystrokes ALPHA 8 (P) (1 + . 06 ÷ 4) ^ (4 ALPHA 4 (T)).	`Plot1 Plot2 Plot3` `\Y1■P(1+.06/4)^(` `4T)` `\Y2=` `\Y3=` `\Y4=` `\Y5=` `\Y6=`
Return to the home screen and input the values $P = 5300$ and $t = 10$ with the keystrokes 5300 STO► ALPHA 8 (P) ALPHA . (:) 10 STO► ALPHA 4 (T) ENTER. Evaluate Y1 at these inputs with VARS ► [Y–VARS] 1 [Function] 1 [Y1] ENTER.	`5300→P:10→T` `                10` `Y1` `       9614.297566`
To evaluate Y1 at other inputs, press 2nd ENTER (ENTRY) twice to recall the storing instruction. Change the values and press ENTER. Then press 2nd ENTER (ENTRY) twice to recall Y1 and press ENTER.	`                10` `Y1` `       9614.297566` `6500→P:8→T` `                 8` `Y1` `       10467.10808`

CAUTION: It is very important to note at this point that while we have previously used X as the input variable when entering functions in the Y= list, we do *not* follow this rule when we *evaluate* functions with more than one input variable. However, realize that we should not graph Y1 nor use the TABLE. If you attempt to graph the current Y1 = $A(P, t)$ or use the table, you will see that the calculator considers this Y1 a constant. (Check it out and see that the graph is a horizontal line at about 10467.1 and that all values in the table are about 10467.1.)

9.2.3 VISUALIZING AND ESTIMATING RATES OF CHANGE OF CROSS SECTIONS

The rate of change of a multivariable function (when evaluated at a specific point) is the slope of the line tangent to the graph of a cross-sectional function at that point. We illustrate this concept in this section and the next using the Missouri farmland cross-section equations for elevation: $E(0.8, n)$ and $E(e, 0.6)$. It would be best to use the unrounded functions that were found in Sections 9.2.1a and 9.2.1b of this *Guide*. However, to illustrate the rate-of-change techniques, we use the rounded functions rather than re-enter all the data.

Enter $-2.5n^2 + 2.497n + 799.490 = E(0.8, n)$ in Y1. Because we are going to graph this function, use X, not n, as the input variable. Press WINDOW and set values such as those shown to the right.	Plot1 Plot2 Plot3 \Y1■-2.5X²+2.497 X+799.490 \Y2= \Y3= \Y4= \Y5= \Y6=	WINDOW Xmin=0 Xmax=1.5 Xscl=1 Ymin=795 Ymax=804 Yscl=1 Xres=1

The window settings used above can be obtained by drawing a scatter plot of the data used to find $E(0.8, n)$ or by looking at the $e = 0.8$ column in Table 9.3 in *Calculus Concepts*. The line tangent to the graph at $n = X = 0.6$ can be drawn from either the home screen or the graphics screen. (See Section 3.2.1b of this *Guide* for an explanation of both methods.) We use the home screen method in this section and the graphics screen method in the next.

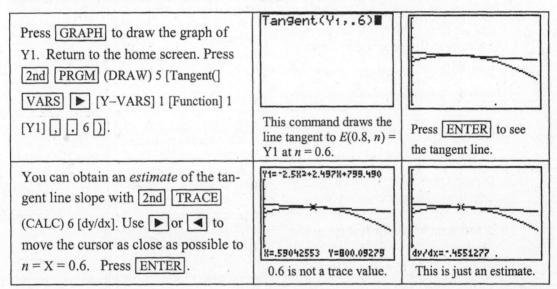

Press GRAPH to draw the graph of Y1. Return to the home screen. Press 2nd PRGM (DRAW) 5 [Tangent(] VARS ▶ [Y-VARS] 1 [Function] 1 [Y1] , . 6).	Tangent(Y1,.6)■ This command draws the line tangent to $E(0.8, n) =$ Y1 at $n = 0.6$.	 Press ENTER to see the tangent line.
You can obtain an *estimate* of the tangent line slope with 2nd TRACE (CALC) 6 [dy/dx]. Use ▶ or ◀ to move the cursor as close as possible to $n = X = 0.6$. Press ENTER.	Y1=-2.5X²+2.497X+799.490 X=.59042553 Y=800.09279 0.6 is not a trace value.	dy/dx=-.4551277 This is just an estimate.

We could have obtained a much better estimate than the one given above if we had found the slope of the tangent line at X = 0.6 rather than a value close to that number. The TI-83 allows you to do this on the home screen or graphically. Here we illustrate the graphical method.

Have $E(0.8, n)$, using X as the input variable, in Y1. Draw the tangent line to $E(0.8, n)$ at $n = 0.6$ by first drawing the graph of the function with GRAPH. With the graph on the screen, press 2nd PRGM (DRAW) 5 [Tangent(] . 6 ENTER.	 X=.6 y=-.503X+800.39

Because this method gives the equation of the line tangent to the graph at the chosen input, you can see that the slope of the tangent, $\dfrac{dE(0.8, n)}{dn}$ at $n = 0.6$, is about -0.503 foot/mile.

9.2.4 FINDING RATES OF CHANGE USING CROSS-SECTIONAL MODELS

The methods described in the last section require that you use X as the input variable to find rates of change. The process of replacing other variables with X can be confusing. You can avoid having to do this replacement when you use the TI-83 numerical derivative on the home screen to evaluate rates of change at specific input values. We illustrate with the cross-sectional function $E(e, 0.6)$ $= -10.124e^3 + 21.347e^2 - 13.972e + 802.809$ feet above sea level that was found in Section 9.2.1b of this *Guide*. We also demonstrate this method with the cross-section $E(0.8, n)$ that was found in Section 9.2.1a and used in the last section of this *Guide*.

NOTE: Whenever possible, you should use the complete equation found by the TI-83 with this method. However, to avoid re-entering the data, we use the rounded functions to illustrate.

Have $E(0.8, n) = -2.5n^2 + 2.497n + 799.490$, using N as the input variable, in Y1. Enter $E(e, 0.6)$, using E as the input variable, in Y2.
 	
(Remember that because we are not using X as the input variable, we should not draw a graph or use the TABLE.)
	```Plot1 Plot2 Plot3```
	```\Y1◼-2.5N²+2.497```
	```N+799.490```
	```\Y2◼-10.124E^3+2```
	```1.347E²-13.972E+```
	```802.809```
	```\Y3=```
	```\Y4=```

NOTE: These functions could be entered on the home screen instead of in the Y= list. We use Y1 and Y2 in the Y= list because they are convenient locations to hold the equations.

Tell the TI-83 the point at which the derivative is to be evaluated by storing the values. Then, on the home screen, find $\dfrac{dE(0.8, n)}{dn}$ evaluated at $n = 0.6$ with the numerical derivative nDeriv.	```.6→N```
	``` .6```
	```.8→E```
	``` .8```
		```.6→N```
		``` .6```
		```.8→E```
		``` .8```
		```nDeriv(Y1,N,.6)```
		``` -.503```

Find $\dfrac{dE(e, 0.6)}{de}$ evaluated at $e = 0.8$ to be about 0.745 foot per mile.
 	
(Always remember to attach units of measure to the numerical values.)
	It is not necessary to again store the values for N and E unless for some reason they have been changed.	``` .6```
		```.8→E```
		``` .8```
		```nDeriv(Y1,N,.6)```
		``` -.503```
		```nDeriv(Y2,E,.8)```
		``` .74510988```

If you want to see the line tangent to the graph of $E(e, 0.6)$ at $e = 0.8$, first change E in Y2 to X and turn off Y1. Draw the graph and use the Tangent instruction in the DRAW menu.	```Plot1 Plot2 Plot3```
	```\Y1=-2.5N²+2.497```
	```N+799.490```
	```\Y2◼-10.124X^3+2```
	```1.347X²-13.972X+```
	```802.809```
	```\Y3=```
	```\Y4=```
		```X=.8```
		```Y=.74510988X+799.513904_```

9.3 Partial Rates of Change

When holding all but one of the input variables in a multivariable function constant, you are actually looking at a function of one input variable. Thus, all of the techniques for finding derivatives that we discussed previously can be used. In particular, the calculator's numerical derivative nDeriv can be used to find partial rates of change at specific values of the varying input variable.

Although your TI-83 does not give formulas for derivatives, you can use it as discussed in Sections 4.3.2b and/or 4.3.2c of this *Guide* to check your answer for the algebraic formula for a partial derivative.

9.3.1 **NUMERICALLY CHECKING PARTIAL DERIVATIVE FORMULAS** As mentioned in Chapter 4, the basic concept in checking your algebraically-found partial derivative formula is that your formula and the calculator's formula computed with nDeriv should have the same outputs when each is evaluated at several different randomly-chosen inputs. You can use the methods in Section 9.2.2 of this *Guide* to evaluate each derivative formula at several different inputs and determine if the same numerical values are obtained from each formula.

We illustrate these ideas by checking the answers for the partial derivative formulas found in parts *b* and *d* of Example 1 in Section 9.3 of *Calculus Concepts* for the following function:

The accumulated value of an investment of *P* dollars over *t* years at an APR of 6% compounded quarterly is $A(P, t) = P(1.061363551^t)$.

Recall that the syntax for the calculator's numerical derivative is

nDeriv(function, symbol for input variable, point at which the derivative is evaluated)

Enter the function *A*, using the letters P and T that appear in the formula, in Y1.
 	
 Part *b* of Example 1 asks for a formula for $\partial A/\partial t$, so enter your formula (which may not be the same as the one that is shown to the right) in Y2. Enter the TI-83's derivative, using T as the changing input, in Y3.
	```Plot1 Plot2 Plot3```
	```\Y1=P(1.06136355```
	```1)^T```
	```\Y2=Pln(1.061363```
	```551)(1.061363551```
	```)^T```
	```\Y3=nDeriv(Y1,T,```
	```T)```

Store a value in T and call up Y2 and Y3. Do this for several different inputs. If the outputs of Y2 and Y3 are the same, your answer in Y2 is *probably* correct.	```5→T:Y2```
	``` 521.3728632```
	```Y3```
	``` 521.3728637```
	```10→T:Y2```
	``` 702.2136524```
	```■``` ```702.2136524```
	```Y3```
	``` 702.2136525```
	```25→T:Y2```
	``` 1715.662291```
	```Y3```
	``` 1715.662292```

NOTE: Remember that 2nd ENTER (ENTRY) recalls previously-entered statements so you do not have to spend time re-entering them. Also recall that ALPHA . (:) joins two or more statements together. Only the value of the last statement is printed, so do not join all three statements shown in the box above together with this symbol.

You may find it more convenient to numerically check your answer using the TABLE. If so, you must remember that when using the TABLE, the TI-83 considers X as the variable that is changing. When finding a partial derivative formula, all other variables are held constant except the one that is changing. So, to use the TABLE (or draw a graph) with a multivariable function, just store values in all constants and call the changing variable X. Then, proceed according to the directions given in Chapter 4. We illustrate this with part *d* of Example 1.

Have the function *A*, using the letters P and T that appear in the formula, in Y1. Because P is the variable that is changing in $\partial A/\partial P$, replace every P with an X. Turn off Y1.
 	
 Enter your formula for $\partial A/\partial P$ (which may not be the same as the one that is shown to the right) in Y2. Enter the TI-83's derivative, using X as the changing input, in Y3.
	```Plot1 Plot2 Plot3```
	```\Y1=X(1.06136355```
	```1)^T```
	```\Y2=(1.061363551```
	```)^T```
	```\Y3=nDeriv(Y1,X,```
	```X)```
	```\Y4=```

Return to the home screen. Because T is constant, enter any reasonable value, except 0, in T. (You could repeat what comes next for T = 2, 10, and 37 as shown in part *c* of Example 1.) (If you need additional instructions on using the TABLE, see Section 1.1.1f of this *Guide*.)	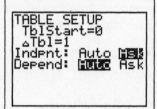

Have TBLSET set to ASK in the Indpnt: location. Go to the TABLE and input some reasonable values for *t* = X. If the values shown for Y2 and Y3 are the same, your formula in Y2 for $\partial A/\partial P$ is *probably* correct.	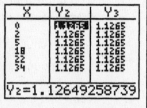

Why are the values in Y2 and Y3 the same for all values of T? It is because the function in Y1 is a linear function with a constant slope. Remember that T is a constant because P is the changing variable in $\partial A/\partial P$.

NOTE: If you have an unrounded function (one fit to data) in Y1 and use the rounded function to compute your partial derivative, the Y2 and Y3 columns will be slightly different.

 ## 9.4 Compensating for Change

As you have just seen, the TI-83 closely estimates numerical values of partial derivatives using its nDeriv function. This technique can also be very beneficial and help you eliminate many potential calculation mistakes when you find the rate of change of one input variable with respect to another input variable (that is, the slope of the tangent line) at a point on a contour curve.

9.4.1a EVALUATING PARTIAL DERIVATIVES OF MULTIVARIABLE FUNCTIONS The last few sections of this *Guide* indicate how to estimate and evaluate partial derivatives using cross-sectional models. The TI-83 evaluates partial derivatives calculated directly from multivariable function formulas using the same procedures. The most important thing to remember is that you must supply the name of the input variable that is changing and the values at which the partial derivative is evaluated. We illustrate using the body-mass index function that is in Example 1 of Section 9.4 of *Calculus Concepts*:

A person's body-mass index is given by $B(h, w) = \dfrac{0.4536w}{0.00064516h^2}$ where *h* is the person's height in inches and *w* is the person's weight in pounds. We first find B_h and B_w at a specific height and weight and then use those values in the next section of this *Guide* to find the value of the derivative $\dfrac{dw}{dh}$ at that particular height and weight. The person in this example is 5 feet 7 inches tall and weighs 129 pounds.

Enter *B* in the Y1 location of the Y= list, using the letters H and W for the input variables.
 	
Next, store the values of H and W at the given point.
	Plot1 Plot2 Plot3
	\Y1█.4536W/(.000
	64516H²)
	\Y2=
	\Y3=
	\Y4=
	\Y5=
	\Y6=
		67→H
		                   67
		129→W
		                  129

CAUTION: The most common mistake made in using this method is forgetting to store the values of the point at which the derivative is to be evaluated. The nDeriv instruction specifies the value at only one of the inputs, and you must tell the calculator the values of all inputs.

Find the value of B_h at $h = 67$ and $w = 129$ by evaluating the expression shown to the right. The symbol B_h (or $\partial B/\partial h$) tells you that h is varying and w is constant, so h is the variable you type in the numerical derivative. Next, enter the value of h. Store this result in N for use in the next section of this *Guide*.	```nDeriv(Y₁,H,67)```
	                -.6031160845
	```Ans→N```
	                -.6031160845

Find the value of B_w at $h = 67$ and $w = 129$ by evaluating the expression shown to the right. The symbol B_w (or $\partial B/\partial w$) tells you that w is varying and h is constant, so w is the variable you type in the numerical derivative. Next, enter the value of w. Store this result in D for use in the next section of this *Guide*.	```nDeriv(Y₁,W,129)```
	                 .156623169
	```Ans→D```
	                 .156623169

9.4.1b FINDING THE SLOPE OF A LINE TANGENT TO A CONTOUR CURVE

We continue the previous illustration with the body-mass index function in Example 1 of Section 9.4 of *Calculus Concepts*. Part *a* of Example 1 asks for $\frac{dw}{dh}$ at the point (67, 129) on the contour curve corresponding to the person's current body-mass index. The formula is $\frac{dw}{dh} = \frac{-B_h}{B_w}$.

An easy way to remember this formula is that whatever variable is in the numerator of the derivative (in this case, w) is the same variable that appears as the changing variable in the denominator of the slope formula. This is why we stored B_w as D (for denominator) and B_h as N (for numerator). Don't forget to put a minus sign in front of the numerator.

In the previous section, we stored B_h as N and B_w as D. So, $$\frac{dw}{dh} = \frac{-B_h}{B_w} = -\text{N} \div \text{D}.$$ (Storing these values also avoids round-off error.) The rate of change is about 3.85 pounds per inch.	```-N/D``` 3.850746274

9.4.1c COMPENSATING FOR CHANGE

When one input of a two-variable multivariable function changes by a small amount, the value of the function is no longer the same as it was before the change. The methods illustrated below show how to determine the amount by which the other input must change so that the output of the function remains at the value it was before any changes were made. We again continue the previous illustration with the body-mass index function and part *b* of Example 1 of Section 9.4 of *Calculus Concepts*.

To estimate the change in weight needed to compensate for growths of 0.5 inch, 1 inch, and 2 inches if the person's body-mass index is to remain constant, find $\Delta w \approx \frac{dw}{dh}(\Delta h)$ at the given values of Δh. Note that the results are changes in weight, so the units that should be attached are *pounds*.	``` 3.850746274```
	```Ans→T```
	``` 3.850746274```
	```T*.5```
	``` 1.925373137```
	```T*2```
	``` 7.701492549```

NOTE: Again, to avoid rounding error, it is easiest to store the slope of the tangent line in some location, say T, and use the unrounded value to calculate (as shown above).

Chapter 10   Analyzing Multivariable Change: Optimization

 ## 10.2 Multivariable Optimization

As you might expect, multivariable optimization techniques that you use with your TI-83 are very similar to those that were discussed in Chapter 5. The basic difference is that the algebra required to get the expression that comes from solving a system of equations with several unknowns reduced to one equation in one unknown is sometimes difficult. However, once your equation is of that form, all the optimization procedures are basically the same as those that were discussed previously.

10.2.1a FINDING CRITICAL POINTS USING ALGEBRA AND THE SOLVER   Critical points for a multivariable function are points at which maxima, minima, or saddle points occur. We begin the process of finding critical points of a smooth, continuous multivariable function by using derivative formulas to find the partial derivative with respect to each input variable and setting these partial derivatives each equal to 0. We next use algebraic methods to obtain one equation in one unknown input. Then, you can use your calculator's solver to obtain the solution to that equation. This method of solution works for all types of equations. We illustrate these ideas with the postal rate function given at the beginning of Section 10.2:

> The volume of a rectangular package that contains the maximum amount of printed material and is sent at the bound printed matter postal rate is given by
>
> $V(h, w) = 108hw - 2h^2w - 2hw^2$ cubic inches
>
> where h inches is the height and w inches is the width of the package.

Find the two partial derivatives and set each of them equal to 0 to obtain these equations:

$$V_h: \quad 108w - 4hw - 2w^2 = 0 \tag{1}$$
$$V_w: \quad 108h - 2h^2 - 4hw = 0 \tag{2}$$

WARNING: Everything that you do with your calculator depends on the partial derivative formulas that you find using derivative rules. Be certain that you check your work before using any of the following solution methods.

Next, solve one of the equations for one of the variables, say equation 2 for w, to obtain

$$w = \frac{108h - 2h^2}{4h}$$

Remember that to *solve* for a quantity means that it must be by itself on one side of the equation without the other side of the equation containing that letter. Now, let your TI-83 work.

Clear the Y= list. Enter the expression for w in Y2. Be certain that you enclose numerators and denominators of fractions in parentheses. (We later enter the function V in Y1.)	Plot1 Plot2 Plot3 \Y1= \Y2■(108H-2H²)/(4H) \Y3= \Y4= \Y5= \Y6=	Plot1 Plot2 Plot3 \Y1= \Y2■(108H-2H²)/(4H) \Y3■108W-4HW-2W² \Y4= \Y5=
Type the left-hand side of the *other* equation (here, equation 1) in Y3. (The expression you type in Y3 must be equal to zero.)	This step tells the TI-83 that w = Y2.	Equations 1 and 2 are now in the TI-83.

Now, replace *every* W in Y3 with the symbol Y2 by placing the cursor on each W location and pressing VARS ► [Y–VARS] 1 [Function] 2 [Y2]. What you have just done is substitute the expression for W from equation 2 in equation 1!

The expression in Y3 is the left-hand side of an equation that equals 0 and it contains only one variable, namely *h*.

```
Plot1 Plot2 Plot3
\Y1=
\Y2◼(108H-2H²)/(
4H)
\Y3◼108Y2-4HY2-2
Y2²
\Y4=
\Y5=
```

The next step is to use the SOLVER to solve the equation Y3 = 0. Try different guesses and see that they all result in the same solution for this particular equation.

```
EQUATION SOLVER
eqn:0=Y3
```

```
Y3=0
•H=18.000000000…
 bound=(-1ᴇ99,1…
•left-rt=0
```

WARNING: You need to closely examine the equation in Y3 and see what type of function it represents. In this case, Y3 contains H to no power higher than one, so it is a linear equation and has only one root. If Y3 contains the variable squared, the equation is quadratic and you need to try different guesses because there could be two solutions, and so forth.

Return to the home screen and press ALPHA ^ [H] ENTER to check that the TI-83 knows that the answer from the SOLVER is a value of H.

Next, find *w* by evaluating Y2. The TI-83 does not know that the output of Y2 equals W, so be sure to store this value in W.

```
H
              18
Y2
              18
Ans→W
              18
```

The last thing to do is to find the output $V(h, w)$ at the current values of *h* and *w*. To do this, enter the function V in Y1 and call up Y1 on the home screen.

```
Plot1 Plot2 Plot3
\Y1◼108HW-2H²W-2
HW²
\Y2◼(108H-2H²)/(
4H)
\Y3◼108Y2-4HY2-2
Y2²
\Y4=
```

```
Y1
          11664
```

The critical point has coordinates $h = 18$ inches, $w = 18$ inches, and $V = 11,664$ cubic inches.

10.2.1b CLASSIFYING CRITICAL POINTS USING THE DETERMINANT TEST   Once you find one or more critical points, the next step is to classify each as a point at which a maximum, a minimum, or a saddle point occurs. The Determinant Test often will give the answer. Also, because this test uses derivatives, the calculator's numerical derivative nDeriv can help.

We illustrate with the critical point that was found in Section 10.2.1a of this *Guide*. To use the Determinant Test, we need to calculate the four second partial derivatives of V and then evaluate them at the critical point values of *h* and *w*.

Enter the functions V in Y1, V_h in Y2 and V_w in Y3. These quantities are given on page A-92 of this *Guide*.

H and W should contain the <u>unrounded</u> values of the inputs at the critical point. (Here, H and W are integers, but this will not always be the case.)

```
Plot1 Plot2 Plot3
\Y1◼108HW-2H²W-2
HW²
\Y2◼108W-4HW-2W²
\Y3◼108H-2H²-4HW
\Y4=
```

```
H
              18
W
              18
```

Take the derivative of Y2 = V_h with respect to h and we have V_{hh}, the partial derivative of V with respect to h and then h again. Find V_{hh} = −72 at the critical point.

```
nDeriv(Y₂,H,H)
               -72
nDeriv(Y₃,H,H)
               -36
```

Take the derivative of Y3 = V_w with respect to h and we have V_{wh}, the partial derivative of V with respect to w and then h. Find V_{wh} = −36 at the critical point.

If you prefer, enter the value of H in the 3rd position.

Take the derivative of Y2 = V_h with respect to w and we have V_{hw}, the partial derivative of V with respect to h and then w. Find V_{hw} = −36 at the critical point. (*Note*: V_{wh} must equal V_{hw}.)

```
nDeriv(Y₂,W,W)
               -36
nDeriv(Y₃,W,W)
               -72
```

Take the derivative of Y3 = V_w with respect to w and we have V_{ww}, the partial derivative of V with respect to w and then w again. Find V_{ww} = −72 at the critical point.

If you prefer, enter the value of W in the 3rd position.

The second partials matrix is $\begin{bmatrix} V_{hh} & V_{hw} \\ V_{wh} & V_{ww} \end{bmatrix}$. Find the value of $D = (-72)(-72) - (-36)^2 = 3888$. Because $D > 0$ and $V_{hh} < 0$ at the critical point, the Determinant Test tells us that $(h, w, V) = (18, 18, 11664)$ is a relative maximum point.

```
-72*-72-(-36)²
               3888
```

NOTE: The values of the second partial derivatives were not very difficult to determine without the calculator in this example. However, with a more complicated function, we strongly suggest using the above methods to provide a check on your analytic work to avoid making simple mistakes. You can also use the function in Y1 to check the derivatives in Y2 and Y3. We use some randomly chosen values of H and W to illustrate this quick check rather than use the values at the critical point, because the critical point values give 0 for Y2 and Y3.

Store different values in H and W and evaluate Y2 at these values. Then evaluate the TI-83's derivative of Y1 with respect to H at these same values.

Evaluate Y3 at these stored values of H and W. Then evaluate the TI-83's derivative of Y1 with respect to W at these same values.

```
8.57→H:19.72→W
               19.72
Y₂
         676.0016
nDeriv(Y₁,H,8.57
)
         676.0016
```

```
8.57→H:19.72→W
               19.72
Y₃
         102.6686
nDeriv(Y₁,W,19.7
2)
         102.6686
```

10.2.2 FINDING CRITICAL POINTS USING MATRICES To find the critical point(s) of a smooth, continuous multivariable function, we first find the partial derivatives with respect to each of the input variables and then set the partial derivatives equal to zero. This gives a *system of equations* that needs to be solved. Here we illustrate solving this system using an orderly array of numbers that is called a *matrix*.

WARNING: The matrix solution method applies only to *linear* systems of equations. That is, the system of equations should not have any variable appearing to a power higher than 1 and should not contain a product of any variables. If the system is not linear, you must use algebraic solution methods to solve the system. Using your TI-83 with the algebraic solution method is illustrated in Section 10.2.1a of this *Guide*.

Consider the cake volume index function in Example 1 of Section 10.2 in *Calculus Concepts*:

$$V(l, t) = -3.1l^2 + 22.4l - 0.1t^2 + 5.3t$$

When l grams of leavening is used and the cake is baked at 177°C for t minutes. The system of equations derived from the partial derivatives of V are

$$V_l = -6.2l + 22.4 = 0$$
$$V_t = -0.2t + 5.3 = 0$$

Even though these 2 linear equations can easily be solved for l and t, we use them to illustrate the matrix solution method. We present two methods of forming the matrix used to solve the system of equations. Regardless of which method you use, you MUST write the equations so that the constant terms are on the right-hand side of the equations and the coefficients of the input variables occupy the same positions on the left-hand side of each equation:

$$-6.2l + \quad 0t = -22.4 \qquad\qquad [1]$$
$$0l + -0.2t = -5.3 \qquad\qquad [2]$$

The matrix menu is accessed with MATRX. You may or may not have numbers (or the same numbers as shown in the first box below) in the column next to the matrix names A, B, etc. The *dimension* of a matrix is the number of rows and columns it contains.

Method 1: Using a Matrix of Coefficients

We choose to use A as the matrix of coefficients. Because there are 2 input variables and 2 equations, we set the dimension of A to 2 by 2; that is, A will have 2 rows and 2 columns.

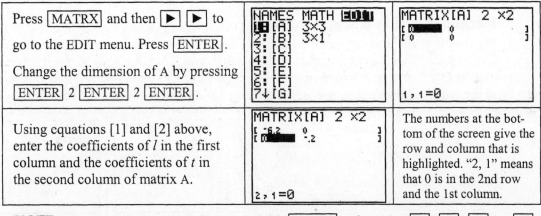

Press MATRX and then ▶ ▶ to go to the EDIT menu. Press ENTER. Change the dimension of A by pressing ENTER 2 ENTER 2 ENTER.	NAMES MATH **EDIT** 1■[A] 3×3 2: [B] 3×1 3: [C] 4: [D] 5: [E] 6: [F] 7↓[G]	MATRIX[A] 2 ×2 [0 0] [0 0] 1,1=0
Using equations [1] and [2] above, enter the coefficients of l in the first column and the coefficients of t in the second column of matrix A.	MATRIX[A] 2 ×2 [-6.2 0] [0 -.2] 2,1=0	The numbers at the bottom of the screen give the row and column that is highlighted. "2, 1" means that 0 is in the 2nd row and the 1st column.

NOTE: You can move around the matrix with ENTER or by using ▶, ◀, ▲, or ▼.

Return to the home screen. We choose to use B as the constant matrix. Follow the same steps as above to set the dimension of B to 2 by 1. Enter the full value of each constant terms.	NAMES MATH **EDIT** 1: [A] 2×2 2■[B] 3: [C] 4: [D] 5: [E] 6: [F] 7↓[G]	MATRIX[B] 2 ×1 [-22.4] [-5.3] 2,1=-5.3
Return to the home screen. The matrix that holds the solution to the system of equations, provided a solution exists, is obtained with MATRX 1 [A] x^{-1} MATRX 2 [B] ENTER.		[A]⁻¹[B] [[3.612903226] [26.5]]

NOTE: Carefully notice the order in which you entered the coefficients. Because the coefficient of l was input first in matrix A, the first value in the solution is l. The other value is t.

We need to use these unrounded values to find the multivariable function output at the critical point. So, store this solution in another matrix, say C, with STO▶ MATRX 3 [C]. Also, enter the multivariable function V in Y1.

We have found that $l \approx 3.61$ grams and $t = 26.5$ minutes. We show how to recall a particular matrix element so that we can use the unrounded values of l and t to find the multivariable function output. Recall that matrix elements are referred to by the row and then the column in which they appear. Because C is a 2 by 1 matrix (2 rows and 1 column) and we set up the equations so that l was the input variable whose coefficient we entered first in matrix A, l is in the (1, 1) position and t is in the (2, 1) position.

Store the value of l into L with MATRX 3 [C] [(1 ,] 1)] STO▶ ALPHA)] (L) ENTER . Store the value of t into T with MATRX 3 [C] [(2 ,] 1)] STO▶ ALPHA 4 (T) ENTER . Find $V(l, t)$ with VARS ▶ [Y–VARS] 1 [Function] 1 [Y1] ENTER .

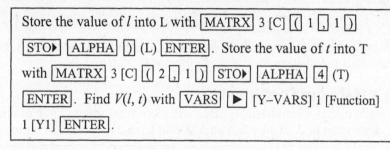

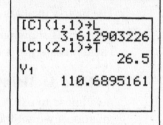

Method 2: Using a Single Matrix

We choose to use A as the single matrix. The cake volume index system of equations contains 2 input variables, a constant term in each equation, and 2 equations, so we set the dimension of A to 2 by 3; that is, A will have 2 rows and 3 columns. (Your TI-83 screens corresponding to the screens shown in the first box below may have numbers different from those indicated.)

Press MATRX and then ▶ ▶ to go to the EDIT menu. Press ENTER .

Change the dimension of A by pressing ENTER 2 ENTER 3 ENTER .

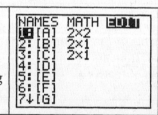

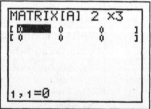

NOTE: You can move around the matrix with ENTER or by using ▶ , ◀ , ▲ , or ▼ .

Using equations [1] and [2] on page A-95, enter the coefficients of l in the first column, the coefficients of t in the second column, and the constant terms in the third column of matrix A.

The numbers at the bottom of the screen give the row and column that is highlighted. "2, 3" means that –5.3 is in the 2nd row and the 3rd column.

Return to the home screen. The keystrokes that give the solution to the system of equations, provided that a solution exists, are MATRX ▶ [MATH] ALPHA MATRX [rref(] MATRX 1 [A])] ENTER . Press and hold ▶ to view all the digits in the third column.

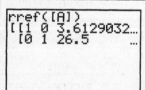

Carefully notice the order in which you entered the coefficients. Because the coefficient of l was input first in matrix A, the first value in the third column of the solution is the value of l. The second value in the third column is the value of t.

| We need to use the unrounded values of $l$ and $t$ to find the multivariable function output at the critical point. So, store this solution in another matrix, say C, with STO▸ MATRX 3 [C]. Also, enter the multivariable function $V$ in Y1. | ```
rref([A])
[[1 0 3.6129032…
 [0 1 26.5    …
Ans→[C]
[[1 0 3.6129032…
 [0 1 26.5    …
``` | ```
Plot1 Plot2 Plot3
\Y1◘-3.1L²+22.4L
-.1T²+5.3T
\Y2=
\Y3=
\Y4=
\Y5=
\Y6=
``` |

We illustrate how to recall a particular matrix element so that we can use the unrounded values of l and t to find the multivariable function output. Recall that a matrix element is referred to by the row and then the column in which it appears. Because C is a 2 by 3 matrix (2 rows and 3 columns) and we set up the equations so that l is the input variable whose coefficient we entered first in matrix A, l is in the (1, 3) position and t is in the (2, 3) position.

| On the home screen, store the value of $l$ into L with MATRX 3 [C] ([ 1 , 3 ]) STO▸ ALPHA ]) (L) ENTER. Store the value of $c$ into C with MATRX 3 [C] ([ 2 , 3 ]) STO▸ ALPHA 4 (T) ENTER. Find $V(l, t)$ with VARS ▶ [Y–VARS] 1 [Function] 1 [Y1] ENTER. | ```
[C](1,3)→L
        3.612903226
[C](2,3)→T
              26.5
Y1
        110.6895161
``` |

10.3 Optimization Under Constraints

Optimization techniques on your calculator when a constraint is involved are the same as the ones discussed in Sections 10.2.1a and 10.2.2 except that there is one additional equation in the system of equations to be solved.

10.3.1 FINDING OPTIMAL POINTS ALGEBRAICALLY AND CLASSIFYING OPTIMAL POINTS UNDER CONSTRAINED OPTIMIZATION

We illustrate solving a constrained optimization problem with the functions given in Example 1 of Section 10.3 – the Cobb-Douglas production function $f(L, K) = 48.1L^{0.6}K^{0.4}$ subject to the constraint $g(L, K) = 8L + K = 98$ where L worker hours (in thousands) and K thousand capital investment are for a mattress manufacturing process.

We first find the critical point(s). Because this function does not yield a linear system of partial derivative equations, we use the algebraic method. We employ a slightly different order of solution than that shown in the text. The system of partial derivative equations is

$$\left.\begin{array}{l} 28.86L^{-0.4}K^{0.4} = 8\lambda \\ 19.24L^{0.6}K^{-0.6} = \lambda \\ 8L + K = 98 \end{array}\right\} \Rightarrow \quad \begin{array}{l} 28.86L^{-0.4}K^{0.4} = 8(19.24L^{0.6}K^{-0.6}) \qquad [4] \\[4pt] K = 98 - 8L \qquad\qquad\qquad\qquad\qquad\quad [5] \end{array}$$

Equation 4 was derived by substituting λ from the second equation on the left into the first, and equation 5 was derived by solving the third equation on the left for K. We now solve this system of 2 equations (equations 4 and 5) in 2 unknowns (L and K) using the methods shown in Section 10.2.1a.
```

Clear the Y= list. Enter the function f in Y1 and the expression for K in Y2. Rewrite the *other* equation (equation 4) so that it equals 0, and enter the non-zero side in Y3. Substitute Y2 into Y3. (See the note below.)	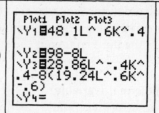	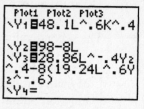

NOTE: Remember that K = Y2. Put the cursor on the first K in Y3 and replace K by Y2. Do the same for the other K in Y3. The expression now in Y3 is the left-hand side of an equation that equals 0 and contains only one variable, namely L. (We are not sure how many answers there are to this equation.)

Use the SOLVER to solve the equation Y3 = 0. Try several different guesses and see that they all result in the same solution. On the home screen, store the result in L and call up Y2 to find K. (Store this value in K.) Enter Y1 to display the value of f at this point.	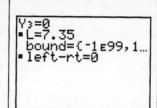

    Classifying critical points when a constraint is involved is done by graphing the constraint on a contour graph or by examining close points. Your TI-83 cannot help with the contour graph classification – it must be done by hand. We illustrate the procedure used to examine close points for this Cobb-Douglas production function.

    We now test close points to see if this output value of f is maximum or minimum. Remember that whatever close points you choose, they must be near the critical point and they must be on the constraint g.

WARNING: Do not round during this procedure. Rounding of intermediate calculations and/or inputs can give a false result when the close point is very near to the optimal point.

Choose a value of L that is less than $L = 7.35$, say 7.3. Find the value of K so that $8L + K = 98$. Remember that K = Y2, so store 7.3 in L and call up Y2. Store this value in K. Then find the value of f at L = 7.3, K = 39.6.
 	
 At this close point, the value of f is less than the value of f at the critical point (690.6084798…).
	``
	7.3→L:Y2
	              39.6
	Ans→K
	              39.6
	Y1
	        690.5845641
	``

Choose another value of L, this time one that is more than $L = 7.35$, say 7.4. Find the value of K so that $8L + K = 98$ by storing 7.4 in L and calling up Y2. Store this value in K. Then find the value of f at L = 7.4, K = 38.8.
 	
 At this close point, the value of f is less than the value of f at the critical point (690.6084798…). Thus, (7.35, 39.2, 690) is a maximum point for mattress production.
	``
	7.4→L:Y2
	              38.8
	Ans→K
	              38.8
	Y1
	        690.5844554
	``

NOTE: Because the constraint is the equation in Y2, when we call up Y2 in the procedure above we are finding points *on* the constraint g. If you have the constraint in a different location, you need to use the constraint location in the Y= list to find the value of K.

10.3.2 FINDING OPTIMAL POINTS USING MATRICES AND CLASSIFYING CRITICAL POINTS UNDER CONSTRAINED OPTIMIZATION Remember that a matrix solution method only can be used with a system of linear partial derivative equations. We choose to illustrate the matrix process of solving with matrix Method 2 (discussed in Section 10.2.2) for the sausage production function in Example 2 of Section 10.3 in *Calculus Concepts*. The system of partial derivative equations and the constraint, written with the constant terms on the right-hand side of the equations and the coefficients of the three input variables in the same positions on the left-hand side of each equation, is

$$-5.83s - \lambda = {}^-1.13$$
$$-5.83w \qquad - \lambda = {}^-1.04$$
$$w \ + \ s \ = 1$$

(The order in which the variables appear does not matter as long as it is the same in each equation.)

When a particular variable does not appear in an equation, its coefficient is zero. Go to the MATRX [EDIT] menu, and set the dimensions of matrix A to be 3 by 4 (that is, 3 rows and 4 columns).

Enter the coefficients of *w* in the first column, the coefficients of *s* in the second column, the coefficients of λ in the third column, and the constant terms in the fourth column of matrix A.	MATRIX[A] 3 ×4
	[0 -5.83 -1
	[-5.83 0 -1
	[1 1 0
	
	3,1=1
		MATRIX[A] 3 ×4
		  -5.83 -1 -1.13]
		   0 -1 -1.04]
		   1 0 1]
		
		3,4=1

Find the solution from the home screen with the instruction rref([A]) in the MATRX MATH menu. The solution to the system of equations is $w \approx 0.492$, $s \approx 0.508$, and $\lambda = {}^-1.83$.
 	
Store this result in matrix C for use with what follows.
		[[1 0 0 .492281…
		  [0 1 0 .507718…
		  [0 0 1 -1.83
		Ans→[C]
		[[1 0 0 .492281…
		  [0 1 0 .507718…
		  [0 0 1 -1.83

(See the discussion of Method 2 on page A-96 of this *Guide* for more detailed instructions.)

Because a contour graph is given (See Figure 10.24 in the text), it is easiest to use it to verify that the optimal point is a minimum. However, we show the method of choosing close points on the constraint to give another example of this method of classification of optimal points.

First, enter the function *P* with inputs *w* and *s* into Y1. Place the constraint, solved for *s*, in Y2. Next, retrieve the unrounded values of *w* and *s* from matrix C and substitute them into the function to find the optimal value.	Plot1 Plot2 Plot3
	\Y1◻10.65+1.13W+
	1.04S-5.83WS
	\Y2◻1-W
	\Y3=
	\Y4=
	\Y5=
	\Y6=
		[C](1,4)→W
		        .4922813036
		[C](2,4)→S
		        .5077186964
		Y1
		        10.27715266

Choose a value of W that is less than 0.492, say 0.48. Store 0.48 in W and call up Y2 to find the value of S. Store this value in *S*. Then find the value of *f* at W = 0.48, K = 0.52.
 	
At this close point, the value of *P* is more than the value of *P* at the critical point.
		.48→W:Y2
		          .52
		Ans→S
		          .52
		Y1
		       10.278032

Now choose a value of W that is more than 0.492, say 0.51. Store 0.51 in W and call up Y2 to find the value of S. Store this value in *S*. Then find the value of *f* at W = 0.48, K = 0.49.
 	
At this close point, the value of *P* is more than the value of *P* at the critical point.
		.51→W:Y2
		          .49
		Ans→S
		          .49
		Y1
		       10.278983

Because the value of *P* at the close points is more than its value at the critical point, $w \approx 0.492$ whey protein, $s \approx 0.508$ skim milk powder, and $P \approx 10.277\%$ cooking loss is a minimum point.

Chapter 11 Dynamics of Change: Differential Equations and Proportionality

11.3 Numerically Estimating by Using Differential Equations: Euler's Method

Many of the differential equations we encounter have solutions that can be found by determining an antiderivative of a given rate-of-change function. Thus, many of the techniques that we learned using the TI-83's numerical integration function apply to this chapter. (See Chapter 6 of this *Guide*.)

11.3.1 EULER'S METHOD FOR A DIFFERENTIAL EQUATION WITH ONE INPUT VARIABLE You may encounter a differential equation that cannot be solved by standard methods and you may need to draw an accumulation graph for a differential equation without first finding an antiderivative. In either of these cases, numerically estimating a solution using Euler's method is helpful. This method relies on the use of the derivative of a function to approximate the change in that function. Recall from Section 5.1 of *Calculus Concepts* that the approximate change in a function f at a point is the rate of change of f at that point times a small change in x. That is,

$$f(x + h) - f(x) \approx f'(x) \cdot h \text{ where } h \text{ represents the small change in } x$$

Now, if we let $b = x + h$ and $x = a$, the above expression becomes

$$f(b) - f(a) \approx f'(a) \cdot (b - a) \quad \text{or} \quad f(b) \approx f(a) + (b - a) \cdot f'(a)$$

The starting values for the coordinates of the point (a, b) will be given to you and are often called the *initial condition*. The next step is to repeatedly apply the formula given above to use the slope of the tangent line at $x = a$ to approximate the change in the function between the inputs a and b. When h, the distance between a and b, is fairly small, Euler's method will often give close numerical estimates of points on the solution to the differential equation containing $f'(x)$.

WARNING: Be wary of the fact that there is some error involved in each step of the Euler approximation process that is compounded when each result is used to obtain the next result.

We illustrate Euler's method for a differential equation containing one input variable with the differential equation in Example 1 of Section 11.3. This equation gives the rate of change of the total sales of a computer product t years after the product was introduced:

$$\frac{dS}{dt} = \frac{6.544}{\ln(t + 1.2)} \text{ billion dollars per year}$$

Because Euler's method involves a repetitive process, a program that performs the calculations used to find the approximate change in the function can save you time and eliminate computational errors and some error in rounding.

Before using this program, you must have the differential equation in the Y1 location of the Y= list with X as the input variable. Access program Euler with PRGM .		

Note that your program list may not be the same as the one shown above. The code for program EULER is listed in the *TI-83 Program Appendix*.

Run the program. Each time the program stops for input or for you to view a result, press ENTER to continue.
 	
We choose to use 16 steps. Enter this value. The interval is 4 years, so enter the step size $= \dfrac{\text{length of interval}}{\text{number of steps}} = \dfrac{4}{16} = 0.25$.
	` HAVE DY/DX IN Y1`
	
	`NUMBER OF STEPS=`
	`16`
	`STEP SIZE= 0.25■`	

The initial condition is given as the point (1, 53.2). Enter these values when prompted for them.	`INITIAL INPUT=`
	`1`
	`INITIAL OUTPUT=`
	`53.2`
	`INPUT,OUTPUT IS`
	`       1.25`
	`   55.27493782`
		The first application of the formula gives an estimate for the value of the total sales at $x = 1.25$:
		$S(1.25) \approx 55.275$

Press ENTER several more times to obtain more estimates for total sales.
 	
Record the input values and output estimates on paper as the program displays them.
	`     57.10065132`
	`INPUT,OUTPUT IS`
	`        1.75`
	`     58.74776643`
	`INPUT,OUTPUT IS`
	`          2`
	`     60.26005352`
		Continue pressing ENTER to obtain more estimates of points on the total sales function S.

When 16 steps have been completed (that is, after the input reaches 5), the program draws a graph of the points *(input, output estimate)* connected with line segments. This solution graph is an estimate of the graph of the differential equation solution.	`     71.70209612`
	`INPUT,OUTPUT IS`
	`         4.75`
	`     72.64207417`
	`INPUT,OUTPUT IS`
	`          5`
	`     73.55942757`
		
		This is an estimate of the graph of the function $S(t)$.

Press Y=, turn Plot3 off, turn Y1 on, and press ZOOM ▲ [Zoomfit] to draw the graph of the differential equation.	`Plot1 Plot2 Plot3`
	`\Y1■6.544/ln(X+1`
	`.2)`
	`\Y2=`
	`\Y3=`
	`\Y4=`
	`\Y5=`
	`\Y6=`
		
		This is the slope graph – the graph of $S'(t)$.

11.3.2 EULER'S METHOD FOR A DIFFERENTIAL EQUATION WITH TWO INPUT VARIABLES

Program EULER can be used when the differential equation is a function of x and y with $y = f(x)$. Follow the same process that is illustrated in the previous section of this *Guide*, but enter $\dfrac{dy}{dx}$ in Y1 using the letters x and y as they are written in the given equation.

If the differential equation is written in terms of variables other than x and y, let the derivative symbol be your guide as to which variable corresponds to the input and which corresponds to the output. For instance, if the rate of change of a quantity is given by $\dfrac{dP}{dn} = 1.346P(1 - n^2)$,

compare $\dfrac{dy}{dx}$ to $\dfrac{dP}{dn}$, entering in Y1 the expression $1.346Y(1 - X^2)$. Use ALPHA 1 (Y) to type Y.

The differential equation may be given in terms of y only. For instance, if $\dfrac{dy}{dx} = k(30 - y)$

where k is a constant, enter Y1 = K(30 – Y). Of course, you need to store a value for k or substitute a value for k in the differential equation before using program EULER. It is always better to store the exact value for a constant instead of using a rounded value.

We illustrate using Euler's method with two input variables by using the equation in Example 2 of Section 11.3 in *Calculus Concepts*.

Enter $\frac{dy}{dx} = 5.9x - 3.2y$ in Y1. Run program EULER. We choose to use 10 steps over an interval of length 2; hence, the step size is 0.2. The initial condition is $y(10) = 50$.	```Plot1 Plot2 Plot3 \Y1■5.9X-3.2Y \Y2= \Y3= \Y4= \Y5= \Y6= \Y7=```	```NUMBER OF STEPS= 10 STEP SIZE= .2 INITIAL INPUT= 10 INITIAL OUTPUT= 50■```
The first estimate for a point on the solution is $y = 29.8$ when $x = 10.2$. Continue until 10 steps are completed; that is, until the input is 12.	```INITIAL INPUT= 10 INITIAL OUTPUT= 50 INPUT,OUTPUT IS 10.2 29.8```	```20.8203948 INPUT,OUTPUT IS 11.8 21.18334213 INPUT,OUTPUT IS 12 21.55000317```
The estimate of $y(12)$ is 21.55. Press ENTER to draw a graph of the Euler estimates. (Note that we cannot graph the differential equation on the TI-83 as we do when there is only one input variable.)		

Troubleshooting the TI-83

❑ CALCULATOR BATTERIES

- *Will I lose all my programs when I change batteries?*

There is no guarantee that changing batteries will not reset your calculator's memory. However, a method that seems to work well is to make sure your calculator is turned off, take the old batteries out one at a time, replacing each with a new battery before removing the next. Be sure you have the direction of the batteries correct (+ and – alternating).

- *What is the backup battery?*

The backup battery is a round battery above the AAA batteries and under a piece of plastic that is fastened to the calculator with a small screw.

- *How long will my batteries last?*

How long your batteries last depends on how much you use your calculator. The four AAA batteries in the calculator usually last for about a year with normal use in a calculus course. The backup battery can last anywhere from 2 to 5 years.

- *When do I need to replace the batteries?*

You need to replace the AAA batteries if your screen is not dark enough and 9 is the value shown in the upper right hand column when you press **2nd**, release that key, and then press and hold the up arrow. (See the second bulleted item under the GENERAL category.)

- *I've replaced the four AAA batteries and my calculator still won't come on. What next?*

Sometimes when you replace batteries, the calculator resets to its lightest screen setting. Darken the display according to the directions given in the second bulleted item under the GENERAL category. If you have replaced the AAA batteries, you have darkened the display, and your calculator still will not turn on, it is likely that the backup battery needs replacing.

❑ DELETING PROGRAMS OR OTHER INFORMATION FROM THE TI-83

- *How do I delete information from my calculator?*

Press **2nd + [MEM] 2 [Delete]**. Next press the number corresponding to what you wish to delete. For instance, to delete a program, press **7 [Prgm]**. Use the down arrow to move the cursor next to the information you want to delete and press **ENTER**. *Be very careful – there is no "undo" command.*

- *The calculator keeps giving me a memory error when I try to enter data or transfer or run a program. What can I do?*

Press **2nd + [MEM] 1 [Check RAM]**. If you have 5000 or less in the MEM FREE location, you need to delete some of the calculator's contents. Pictures (PIC) and graph databases (GDB) take up a lot of room and can be deleted by following the directions in the answer directly above. Game programs also take up a lot of room and create some of the pictures and databases that may be stored in your calculator.

❑ GENERAL INFORMATION

- *Why do my calculator screens not look like the ones in this Guide?*

Make sure your calculator settings are as specified in the *Setup Instructions* on page *A-1*.

- *How do I make the display screen darker or lighter?*

To make the display darker, press **2nd**, release that key, and then press and hold the up arrow until the display is dark enough. To make the display lighter, press **2nd**, release that key, and then press and hold the down arrow until the display is light enough. You should see a number in the upper right corner of the display screen that varies between 0 (light) to 9 (dark.)

- *The calculator doesn't recognize what I have entered in Y1 as the function.*

Anytime you refer to a function in the Y= list, you must type the function name using **VARS right arrow [Y-VARS] 1 [Function]** followed by the number of the desired function location. The TI-83 does not recognize **ALPHA 1 (Y)** as the name of a function in the Y= list.

- *I have either lost or never got the owner's manual for the TI-83. What can I do?*

The *TI-83Guidebook* is available at the TI web site with address **www.TI.com/calculators**.

- *I can't find the information I need in this Guide.*

See the material entitled **How to Use This Guide** following the *Contents* on page vi.

❑ GRAPHING

- *Why don't I see a graph when I press the correct keys to make it draw?*

Make sure that you have the function entered in the graphing (Y=) list, using X for the input variable. (See Sections 1.1.1a, b.) Also check that your function is turned on. (See Section 1.2.2b.) If the function is turned on but you still cannot see the graph, check your window settings – maybe the function did graph, but it graphed outside your window.

- *What do I do if a get a strange-looking graph or no graph instead of a scatter plot of data when I press* **ZOOM 9 [ZoomStat]**?

The scatter plot setup has somehow been changed, is not correctly set, or is not turned on. Refer to Section 1.4.2a of this *Guide* for instructions.

- *Do I have to clear the function location in the Y= list before pasting in another function when finding an equation of best fit?*

Most of the time it is not necessary to first clear any previously-entered function from the chosen location of the Y= list. However, if you receive an error message when finding the equation, clear the desired function location and press **2nd ENTER (ENTRY) ENTER** to recall and repeat the regression instruction. If you still obtain an error message, reset the statistical setup as described on the first page of this *Guide*.

- *While trying to draw the graph of an equation that I entered in the Y= list, I get a STAT-PLOT error. What do I do?*

The TI-83 will sometimes not draw the graph of a function when one of the scatter plots is turned on. To correct this problem, choose **1: [Quit]** and press **2nd Y= (STAT PLOT) 4 [Plots Off] ENTER**. The graph should now draw with out any problems.

- *What causes the error message* ERR: WINDOW RANGE *and how do I eliminate this error?*

This error message is usually caused when Ymin = Ymax in the WINDOW settings. The code in a program may be such that the height of the window (Ymin – Ymax) is set to zero. If you attempt to draw a scatter plot when the output is constant, this error results because the TI-83 uses the output data to set the range for the WINDOW. To correct this error, manually set Ymin and Ymax and press **GRAPH** to draw the graph.

❏ LISTS

- *I don't have all the lists when I press **STAT ENTER**. (For instance, what do I do if list L1 is missing?)*

Make sure your calculator settings are as specified in Figure 2 using the second bulleted item in the list describing the Setup Instructions on the first page of this *Guide*.

❏ PROGRAMS

- *Where can I get the programs that are used in this Guide?*

The programs can be transferred to your calculator from another student or your teacher's calculator and a link chord. If no one has the programs, they can be downloaded from the *Calculus Concepts* CD-ROM or web site.

A TI Graph-Link™ cable and the linking software for your particular calculator are needed to transfer the programs from the Internet to a computer. A special cable is needed to transfer the programs from the computer to your calculator. The linking software is free at the TI web site with address **www.ti.com/calculators**. If you cannot find a cable, ask at the learning center for your school, check with your instructor, or search the TI web site for instructions on how to purchase one.

- *Why won't a program run, or why do I get an error when I try to run a program?*

You may have unknowingly deleted or altered one or more lines of the program code. The easiest method of fixing the problem is to use the link chord to re-transfer the entire program into the calculator. If this is not possible, you can find the correct program code in the *TI-83 Program Appendix* in this *Guide*. Press **PRGM right arrow [EDIT]** and **down arrow** until the name of the corrupt program is highlighted. Press **ENTER**. Compare the code in your calculator with what is printed in the *TI-83 Program Appendix* until you find the error and correct it by retyping the proper code. Consult the *TI-83 Guidebook* that came with your calculator for the location of the symbols in the program code.

- *Program NUMINTGL keeps giving an error when I try to run it. What can I do?*

Delete all functions from the Y= list (except the function in Y1) before using the program named NUMINTGL. This program uses every memory location in the calculator. If you receive an error while running this program, you may have a picture, another program, or something else stored as a single-letter name. For instance, if program NUMINTGL is trying to store a number in memory location T and you have a program called T, the calculator stops and may not give an error message. Delete or rename any programs, pictures, or whatever you have called by a single-letter name before continuing. (See DELETING PROGRAMS OR OTHER INFORMATION FROM THE TI-83 on page A-103 of this *Guide*.)

TI-83 Program Appendix

The programs listed below are referenced in Part *A* of the *Graphing Calculator Instruction Guide* for *Calculus Concepts*. They should be transferred to your TI-83 via a cable by using the LINK mode and another TI-83 calculator or by using the TI-GRAPH LINK™ cable and GRAPH LINK™ software for a PC or Macintosh computer that is accessible through the site **www.education.ti.com**. These programs are available for download on the *Calculus Concepts* CD-ROM and web site. As a last resort, the programs can be typed into your calculator. Please refer to your *TI-83 Owner's Guidebook* for instructions on entering the programs or transferring them via cable from another TI-82 or TI-83. The program code follows.

Program Name	Program Size (bytes)	Chapter first referenced
DIFF	446	1
NUMINTGL	984	6
EULER	237	11
LSLINE*	473	1
SECTAN*	449	3

*These programs are more for instructional exploration than for use in working problems.

DIFF   • Program
```
ClrHome
dim(L1)→N:N-1→dim(L6)
For(H,1,N-1,1)
L1(H+1)-L1(H)→L6(H)
End
For(H,1,N-2,1)
If L6(H+1)≠L6(H)
Goto 2
End
dim(L2)→M:M-1→dim(L3)
For(A,1,M-1,1)
L2(A+1)-L2(A)→L3(A)
End
M-2→dim(L4)
For(B,1,M-2,1)
L3(B+1)-L3(B)→L4(B)
End
Disp "HAVE X IN L1"
Disp "HAVE Y IN L2-SEE"
Disp "1ST DIFF IN L3,
Disp "2ND DIFF IN L4,"
M-1→dim(L5)
1→E
For(E,1,M,1)
If L2(E)=0
Goto 1:End
For(E,1,M-1,1)
(L3(E)/L2(E))*100→L5(E)
End
Disp "PERCENT CHANGE"
Disp "IN L5"
Stop
Lbl 1
ClrList L5
Disp "PERCENT CHANGE"
```

(Program DIFF *continued)*
```
Disp "NOT CALCULATED"
Stop
Lbl 2
Disp "INPUT VALUES NOT"
Disp "EVENLY SPACED"
Stop
```

EULER   • Program
```
ClrHome
ClrList L1,L2
FnOff
Disp "HAVE DY/DX IN Y1"
Disp ""
Input "NUMBER OF STEPS= ",N
Input "STEP SIZE= ",H
Input "INITIAL INPUT=   ",X
Input "INITIAL OUTPUT= ",Y
For(I,1,N,1)
X→L1(I)
Y→L2(I)
Y1→T
X+H→X
Y+H*T→Y
Disp "INPUT,OUTPUT IS"
Disp X
Disp Y
Pause
End
X→L1(N+1):Y→L2(N+1)
Plot3(xyLine,L1,L2,▫)
ZoomStat
```

NUMINTGL   • Program

```
0→A:0→L:0→B
ClrHome
PlotsOff
Disp "ENTER F(X) IN Y1"
Disp ""
Disp "CONTINUE?"
Input "YES(1) NO(2) ",G
If G=2:Stop
Disp ""
Disp "DRAW PICTURES?"
Input "YES(1) NO(2) ",H
ClrHome
Input "LEFT ENDPOINT? ",A
Input "RIGHT ENDPOINT? ",B
If H=1:Then
A→Xmin:B→Xmax
iPart((B-A)/20)→W
If W=0:Goto Z
Lbl U
seq(X,X,A,B,W)→L5
Y1(L5)→L6
min(L6)→Ymin
If Ymin>0:0→Ymin
max(L6)→Ymax
If Ymax<0:0→Ymax
W→Xscl
iPart(abs(Ymax-Ymin)/10)→Yscl
ClrList L5,L6
End
Lbl 0
ClrHome
Disp "ENTER CHOICE:"
Disp "LEFT RECT  (1)"
Disp "RIGHT RECT (2)"
Disp "TRAPEZOIDS (3)
Input "MIDPT RECT (4) ",R
Lbl 1
ClrDraw
Input "N? ",N
(B-A)/N→W
0→S:1→C
Lbl 2
If R=1:Goto 3
If R=2:Goto 4
If R=3:Goto 3
If R=4:Goto 5
Lbl 3
A+(C-1)W→X
X→J:X+W→L
Goto 7
Lbl 4
A+CW→X
X-W→J:X→L
Goto 7
Lbl 5
If H≠1:Then
If N>5:Then
1→Z:W/2→H:A→X
Lbl 8
X+H→X:Y1+S→S
```

(Program NUMINTGL *continued)*

```
A+ZW→X
IS>(Z,N):Goto 8
SW→S:Goto T
End:End
A+CW-W/2→X
X-W/2→J
X+W/2→L
Goto 7
A→G:G+W→G:G→V
Lbl 9
V→X:Y1→Y:V+W→X:4Y+2Y1+S→S
V+2W→V
If V<B:Goto 9
G-W→X:Y1→E
B→X:Y1→F
(W/3)(S+E-F)→S
Goto T
Lbl 7
Y1→K:K+S→S
If H=1:Goto D
Lbl I
IS>(C,N):Goto 2
If R=3:Then
A→X:Y1→P
B→X:Y1→Q
S+(Q-P)/2→S
End
W*S→S
Lbl T
Disp "SUM=",S
Pause:ClrHome
Lbl E
Menu("ENTER CHOICE","CHANGE
N",1,"CHANGE METHOD",0,"QUIT",F)
Lbl F
Stop
Lbl D
If R=3:Then
Y1(L)→M
Else:K→M
End
Line(J,0,J,K)
Line(J,K,L,M)
Line(L,M,L,0)
If C=N:Pause
Goto I
Lbl Z
If B-A≤9:Then
.1→W:Else
1→W:End
Goto U
```

LSLINE   • Program

```
0→A:0→B:1→C
"AX+B"→Y₁
Ymax-Ymin→H
.2H+Ymax→Ymax
FnOff
Text(0,0,"X TICK=",Xscl,"   Y
TICK=",Yscl)
Pause
ClrHome
Lbl 1
Text(0,0,"GUESS  SLOPE,  Y-INTERCEPT")
Pause
FnOn
Input "SLOPE=",A
Input "Y-INTERCEPT=",B
2-Var Stats
Lbl 2
0→S
For(K,1,n)
L₁(K)→X
(L₂(K)-Y₁)² +S→S
Line(L₁(K),L₂(K),X,Y₁)
End
Pause
Disp ""
Disp "SSE=",S
Pause
If C=2
Goto 3
Input "TRY AGAIN? 1Y 2N",C
If C=1
Goto 1
LinReg(ax+b)
"aX+b"→Y₂
Disp ""
Disp "PRESS ENTER TO"
Disp "SEE YOUR LINE"
Disp "AND BESTFIT LINE"
Pause
DispGraph
Pause
ClrHome
Disp "NOW,PRESS ENTER"
Disp "TO SEE ERRORS"
Disp "FOR BESTFIT LINE"
Pause
a→A:b→B
Goto 2
Lbl 3
Disp "SLOPE=",a
Disp "Y-INTERCEPT=",b
FnOff
```

SECTAN   • Program

```
ClrHome
PlotsOff
ClrDraw:2→R
Disp ""
Disp "HAVE F(X)IN Y₁ AND"
Disp "DRAW GRAPH OF F"
Disp ""
Disp "CONTINUE? "
Input "YES(1) NO(2)",C
If C=2:Stop
Disp ""
Disp "X-VALUE OF POINT"
Input "OF TANGENCY? ",A
Lbl 1
Disp ""
Disp "PRESS ENTER TO "
Disp "SEE SECANT LINES"
If R=1:Goto 2
Disp "FROM THE LEFT"
Goto 3
Lbl 2
ClrDraw
Disp "FROM THE RIGHT"
Lbl 3
Disp "APPROACH TANGENT"
Disp "LINE"
Pause
(Xmax-Xmin)/3→K
If K>50:48→K
For(J,1,5,1)
A-K→X
If R=1:A+K→X
(Y₁(X)-Y₁(A))/(X-A)→M
DrawF (M(X-A)+Y₁(A))
K/2→K
End
Pause
If R=1:Goto 4
1→R:Goto 1
Lbl 4
ClrHome
Disp "PRESS ENTER TO"
Disp "SEE TANGENT LINE"
Pause
ClrDraw
Tangent(Y₁,A)
```

Part B

Guide for Texas Instruments TI-86 Graphing Calculator

This *Guide* is designed to offer step-by-step instruction for using your TI-86 graphing calculator with the third edition of *Calculus Concepts: An Informal Approach to the Mathematics of Change*. A technology icon next to a particular example or discussion in the text directs you to a specific portion of this *Guide*. You should also utilize the table of contents in this *Guide* to find specific topics on which you need instruction.

Setup Instructions

Before you begin, check the TI-86 setup and be sure the settings described below are chosen. Whenever you use this *Guide*, we assume (unless instructed otherwise) that your calculator settings are as shown in Figures 1, 2, and 3.

- Press 2nd MORE (MODE) and choose the settings shown in Figure 1 for the basic setup.

- Check the window format by pressing GRAPH MORE F3 (FORMT) and choose the settings shown in Figure 2.

  - If you do not have the darkened choices shown in Figure 1 and Figure 2, use the arrow keys to move the blinking cursor over the setting you want to choose and press ENTER.

  - Return to the home screen with EXIT or 2nd EXIT (QUIT). Note that EXIT EXIT clears the menus from the bottom of the screen.

- Specify the statistical setup as shown in Figure 3 by pressing 2nd − (LIST) F5 [OPS] MORE MORE MORE F3 [SetLE] ALPHA 7 (L) 2 , ALPHA 7 (L) 2 , ALPHA 7 (L) 3 , ALPHA 7 (L) 4 , ALPHA 7 (L) 5 , ENTER. You need this setup for some of the programs referred to in this *Guide* to execute properly.

TI-86 Basic Setup

Figure 1

TI-86 Window Setup

Figure 2

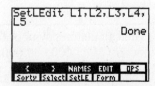

TI-86 Statistical Setup

Figure 3

Basic Operation

You should be familiar with the basic operation of your calculator. With your calculator in hand, go through each of the following.

1. **CALCULATING**   You can type in lengthy expressions; just make sure that you use parentheses when you are not sure of the calculator's order of operations. Always enclose in parentheses any numerators and denominators of fractions and powers that consist of more than one term.

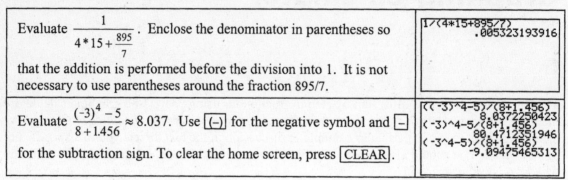

Evaluate $\dfrac{1}{4*15+\dfrac{895}{7}}$. Enclose the denominator in parentheses so that the addition is performed before the division into 1. It is not necessary to use parentheses around the fraction 895/7.

Evaluate $\dfrac{(-3)^4-5}{8+1.456} \approx 8.037$. Use $\boxed{(-)}$ for the negative symbol and $\boxed{-}$ for the subtraction sign. To clear the home screen, press $\boxed{\text{CLEAR}}$.

NOTE: The numerator and denominator must be enclosed in parentheses and $-3^4 \neq (-3)^4$.

Now, evaluate $e^3*0.027$ and $e^{3*0.027}$. Type e^ with $\boxed{\text{2nd}}$ $\boxed{\text{LN}}$ (e^x). The TI-86 will assume you mean $e^3*0.027$ unless you use parentheses around the two values in the exponent to indicate $e^{3*0.027}$.

2. **USING THE ANS MEMORY**   Instead of again typing an expression that was just evaluated, use the answer memory by pressing $\boxed{\text{2nd}}$ $\boxed{(-)}$ (ANS).

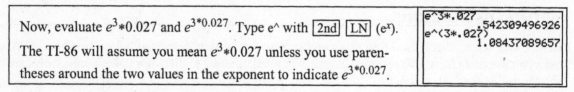

Calculate $\left(\dfrac{1}{4*15+\dfrac{895}{7}}\right)^{-1}$ using this nice shortcut.

Type Ans^{-1} by pressing $\boxed{\text{2nd}}$ $\boxed{(-)}$ (ANS) $\boxed{\text{2nd}}$ $\boxed{\text{EE}}$ (x^{-1}).

3. **ANSWER DISPLAY**   When the denominator of a fraction has no more than three digits, your calculator can provide the answer in the form of a fraction. When an answer is very large or very small, the calculator displays the result in scientific notation.

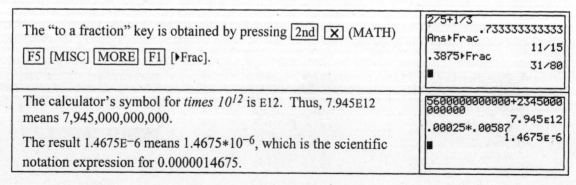

The "to a fraction" key is obtained by pressing $\boxed{\text{2nd}}$ $\boxed{\text{X}}$ (MATH) $\boxed{\text{F5}}$ [MISC] $\boxed{\text{MORE}}$ $\boxed{\text{F1}}$ [▶Frac].

The calculator's symbol for *times 10^{12}* is E12. Thus, 7.945E12 means 7,945,000,000,000.

The result 1.4675E$^-$6 means 1.4675*10^{-6}, which is the scientific notation expression for 0.0000014675.

4. **STORING VALUES**   It may be beneficial to store numbers or expressions for later recall. To store a number, type it, press [STO▶] (note that the cursor changes to the alphabetic cursor [A]), press the key corresponding to the capital letter(s) naming the storage location, and then press [ENTER]. To join several short commands together, use [2nd] [.] (:) between the statements. Note that when you join statements with a colon, only the value of the last statement is shown.

WARNING: The [STO▶] key locks the upper-case alphabetic cursor and [ALPHA] unlocks it. Always look at the screen when you are typing to be certain that you are not entering numbers when you intend to type letters and vice-versa.

Store 5 in *A* and 3 in *B*, and then calculate $4A - 2B$. (Press [ALPHA] to return to the regular cursor.) To recall a value, press [ALPHA], type the letter in which the value is stored, and then press [ENTER].	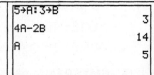

- Storage location names on the TI-86 can be from one to eight characters long, but they must begin with a letter. You cannot name what you are storing with the exact name the TI-86 already uses for a built-in variable (such as LOG).

- Whatever you store in a particular memory location stays there until it is replaced by something else either by you or by executing a program containing that name.

NOTE: The TI-86 allows you to enter upper and lower case letters, and it distinguishes between them. For instance, VOL, Vol, VOl, vol, voL, and so forth are all different names to the TI-86. To type a lower case letter, press [2nd] [ALPHA] before pressing a letter key (note that the cursor changes to [a]). If you cannot remember which combination of upper- and lower-case letters you used for a name, press [2nd] [CUSTOM] [CATLG-VARS] [F2] [ALL] and then press the key of the first letter of the name. Use [▼] to move the cursor next to the name and then press [ENTER].

5. **ERROR MESSAGES**   When your input is incorrect, the TI-86 displays an error message.

If you have more than one command on a line without the commands separated by a colon (:), an error message results when you press [ENTER].	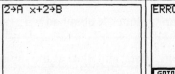	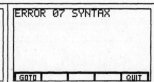

- When you get an error message, press [F1] [Goto] to position the cursor to where the error occurred so that you can correct the mistake or choose [F5] [Quit] to begin a new line on the home screen. When you are executing a program, you should always choose the 1: Quit option upon receiving an error message. Choosing 2: Goto will call up the program code, and you may inadvertently change the program so that it will not properly execute.

A common mistake is using the negative symbol [(−)] instead of the subtraction sign [−] or vice-versa. The TI-86 does *not* give an error message, but a wrong answer results. The negative sign is shorter and raised slightly more than the subtraction sign.	2*⁻3 ⁻6 2−3 ⁻6 2−3 ⁻1

Chapter 1   Ingredients of Change: Functions and Linear Models

 1.1 Models, Functions, and Graphs

Graphing a function in an appropriate viewing window is one of the many uses for a function that is entered in the calculator's graphing list. Because you must enter a function formula on one line (that is, you cannot write fractions and exponents the same way you do on paper), it is very important to have a good understanding of the calculator's order of operations and to use parentheses whenever they are needed.

NOTE: If you are not familiar with the basic operation of the TI-83, you should work through pages A-1 through A-3 of this *Guide* before proceeding with this material.

1.1.1a ENTERING AN EQUATION IN THE GRAPHING LIST The graphing list contains space for 99 equations (if memory is available), and the output variables are called by the names y1, y2, ..., y99. To graph an equation, enter it in the graphing list. You must use x as the input variable if you intend to draw the graph of the equation or use the TI-86 table. We illustrate graphing using the equation in Example 3 of Section 1.1: $v(t) = 3.622(1.093^t)$.

<table>
<tr>
<td>

Press $\boxed{\text{GRAPH}}$ $\boxed{\text{F1}}$ [y(x)=] to access the graphing list.

If there are any previously entered equations that you will no longer use, delete them from the graphing list with $\boxed{\text{F4}}$ [DELf].

</td>
<td>

To clear, but not delete, the location of an equation in the y(x)= list, position the cursor on the line with the equation and press $\boxed{\text{CLEAR}}$.

</td>
</tr>
<tr>
<td>

For convenience, we use the first, or y1, location in the list. We intend to graph this equation, so the input variable must be called x, not *t*. Enter the right-hand side of the equation, $3.622(1.093^x)$, with $3 \boxed{.} 6 2 2 \boxed{(} 1 \boxed{.} 0 9 3 \boxed{\wedge} \boxed{\text{x-VAR}} \boxed{)}$

You should use the $\boxed{\text{x-VAR}}$ key for *x*, not the times sign key, $\boxed{\times}$, nor the capital letter X obtained with $\boxed{\text{ALPHA}}$ $\boxed{+}$ (X).

</td>
<td>

```
Plot1 Plot2 Plot3
\y1■3.622(1.093^x)



Y(X)= WIND ZOOM TRACE GRAPH
 x   y   INSf DELf SELCT▶
```

</td>
</tr>
</table>

CAUTION: Plot1, Plot2, and Plot3 at the top of the y(x)= list should *not* be darkened when you are graphing an equation and not graphing data points. If any of these is darkened, use $\boxed{\blacktriangle}$ until you are on the darkened plot name. Press $\boxed{\text{ENTER}}$ to make the name(s) not dark (that is, to *deselect* the plot). If you do not do this, you may receive a STAT PLOT error message.

1.1.1b DRAWING A GRAPH As is the case with most applied problems in *Calculus Concepts*, the problem description indicates the valid input interval. Consider Example 3 of Section 1.1:

> The value of a piece of property between 1985 and 2005 is given by $v(t) = 3.622(1.093^t)$ thousand dollars where *t* is the number of years since the end of 1985.

The input interval is 1985 ($t = 0$) to 2005 ($t = 20$). Before drawing the graph of *v* on this interval, enter the $v(t)$ equation in the y(x)= list using *x* as the input variable. (See Section 1.1.1a of this *Guide*.) We now draw the graph of the function *v* for *x* between 0 and 20.

> Press [GRAPH] [F2] [WIND] to set the view for the graph. Enter 0 for xMin and 20 for xMax. (For 10 tick marks between 0 and 20, enter 2 for Xscl. If you want 20 tick marks, enter 1 for xScl, etc. xScl does not affect the shape of the graph.)
>
> ```
> WINDOW
>  xMin=0
>  xMax=20
>  xScl=2
>  yMin=-6
>  yMax=6
> ↓yScl=1
> y(x)= WIND ZOOM TRACE GRAPH
> ```

xMin and xMax are, respectively, the settings of the left and right edges of the viewing screen, and yMin and yMax are, respectively, the settings for the lower and upper edges of the viewing screen. xScl and yScl set the spacing between the tick marks on the x- and y-axes. (Leave xRes set to 1 for all applications in this *Guide*.) We now set the values to determine the output view.

> To have the TI-86 determine the output view, press [F3] [ZOOM] [MORE] [F1] [ZFIT] [ENTER].

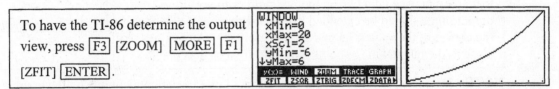

Note that any vertical line drawn on this graph intersects it in only one point, so the graph does represent a function. ([CLEAR] removes the menu from the bottom of the screen if you want to see the entire graphics screen.)

> Press [GRAPH] [F2] [WIND] to see the view set by ZFIT. The view has $0 \le x \le 20$ and $3.622 \le y \le 21.446\ldots$. (Note that ZFIT did not change the x-values you manually set.)
>
> ```
> WINDOW
>  xMin=0
>  xMax=20
>  xScl=2
>  yMin=3.622
>  yMax=21.4462658366
> ↓yScl=1
> y(x)= WIND ZOOM TRACE GRAPH
> ```

We just saw how to have the TI-86 set the view for the output variable. Whenever you draw a graph, you can also manually set or change the view for the output variable.

1.1.1c MANUALLY CHANGING THE VIEW OF A GRAPH   We just saw how to have the TI-86 set the view for the output variable. Whenever you draw a graph, you can also manually set or change the view for the output variable. If for some reason you do not have an acceptable view of a graph or if you do not see a graph, change the view for the output variable with one of the ZOOM options or manually set the window until you see a good graph. (We will later discuss other ZOOM options.) We continue using the function v that is given in Example 3 of Section 1.1, but here assume that you have not yet drawn the graph of v.

> Press [GRAPH] [F2] [WIND], enter 0 for xMin and 20 for xMax, and (assuming we do not know what is the vertical view), enter some arbitrary values for yMin and yMax. (The graph to the right was drawn with yMin = −5 and yMax = 5.) Press [F5] [GRAPH].

> **Evaluating Outputs on the Graphics Screen:** First, press [F4] [TRACE]. Recall we are given that the input is between 0 and 20. If you now type the number that you want to substitute in the function whose graph is drawn, say 0, you see the screen to the right. A 1 appears at the top of the screen because the equation of the function whose graph you are drawing is in y1.

> Press [ENTER] and the input value is substituted in the function. The input and output values are shown at the bottom of the screen. (This method works even if you do not see any of the graph on the screen.)

Substitute the right endpoint of the input interval into the function by pressing 20 ENTER. We see that two points on this function are approximately (0, 3.622) and (20, 21.446).

x=20   y=21.446265837

Press GRAPH F2 [WIND], enter 3.5 for yMin and 22 for yMax, and press F5 [GRAPH]. If the graph you obtain is not a good view of the function, repeat the above process using x-values that are in between the two endpoints in order to see if the output range should be extended in either direction. (Note that the choice of the values 3.5 and 22 was arbitrary. Any values close to the outputs in the points you find are also acceptable.)

y(x)= WIND ZOOM TRACE GRAPH

NOTE: Instead of using TRACE with the exact input to evaluate outputs on the graphics screen, you could use the TI-86 TABLE or evaluate the function at 0 and 20 on the home screen. We next discuss using these features.

1.1.1d TRACING TO ESTIMATE OUTPUTS

You can display the coordinates of certain points on the graph by tracing. Unlike the substitution feature of TRACE that was discussed on the previous page, the x-values that you see when tracing the graph depend on the horizontal view that you choose. The output values that are displayed at the bottom of the screen are calculated by substituting the x-values into the equation that is being graphed. We again assume that you have the function $v(x) = 3.622(1.093^x)$ entered in the y1 location of the y(x)= list.

With the graph on the screen, press F4 [TRACE], press and hold ▶ to move the trace cursor to the right, and press and hold ◀ to move the trace cursor to the left.

x=12.380952381   .y=10.891879602

- Again note that the number corresponding to the location (in the y(x)= list) of the equation that you are tracing appears in the top right corner of the graphics screen.

Trace past one edge of the screen and notice that even though you cannot see the trace cursor, the x- and y-values of points on the line are still displayed at the bottom of the screen. Also note that the graph scrolls to the left or right as you move the cursor past the edge of the current viewing screen.

x=23.650793651   y=29.671938967

Use either ▶ or ◀ to move the cursor near x = 15. We estimate that y is *approximately* 13.8 when x is *about* 15.

It is important to realize that trace outputs should *never* be given as answers to a problem unless the displayed x-value is *identically* the same as the value of the input variable.

x=15.079365079   y=13.845733478

1.1.1e EVALUATING OUTPUTS ON THE HOME SCREEN

The input values used in the evaluation process are *actual* values, not estimated values such as those that are generally obtained by tracing near a certain value. Actual values are also obtained when you evaluate outputs from the graphing screen using the process that was discussed in Section 1.1.3 of this *Guide*.

We again consider the function $v(t) = 3.622(1.093^t)$ that is in Example 3 of Section 1.1.

> Using x as the input variable, enter y1 = 3.622(1.093^ x). Return to the home screen by pressing [2nd] [EXIT] (QUIT). Substitute 15 into the function with [2nd] [ALPHA] (alpha) 0 (Y) 1 [(] 15 [)] Find the value by pressing [ENTER].

```
y1(15)
            13.7483593553
```

NOTE: You do not have to have the *closing* parenthesis on the right if nothing else follows it. To choose another graphing list location, say y2, just type the number corresponding to that function's location, 2, following the lower-case y.

WARNING: You must use a lower case, not upper case, y in order for the TI-86 to recognize the function in the graphing list. You must also use a lower-case x for the *x*-variable.

To now evaluate the function at other inputs, first recall the previous entry with [2nd] [ENTER] (ENTRY). Then edit the expression to the new value.

> For instance, press [2nd] [ENTER] (ENTRY), change 15 to 20 by pressing [◄] [◄] [◄] and typing 20, and then press [ENTER] to evaluate the function at $x = 20$. Evaluate y1 at $x = 0$ by recalling the previous entry with [2nd] [ENTER] (ENTRY), change 20 to 0 with [◄] [◄] [◄] [DEL], and then press [ENTER].

```
y1(20)
            21.4462658366
y1(0)
                    3.622
```

1.1.1f EVALUATING OUTPUTS USING THE TABLE   Function outputs can be determined by evaluating on the graphics screen, as discussed in Section 1.1.1c, or by evaluating on the home screen as discussed in Section 1.1.1e of this *Guide*. You can also evaluate functions using the TI-86 TABLE. When you use the table, you can either enter specific input values and find the outputs or generate a list of input and output values in which the inputs begin with a value called TblStart and differ by a value called ΔTbl.

Let's use the TABLE to evaluate the function $v(t) = 3.622(1.093^t)$ at $t = 15$. Even though you can use any of the function locations, we again choose to use y1. Press [GRAPH] [F1] [y(x)=], clear the function locations, and enter 3.622(1 .093^ x) in location y1 of the y(x)= list.

> After entering the function v in y1, choose the TABLE SETUP menu by pressing [TABLE] [F2] [TBLST]. To generate a list of values beginning with 13 such that the table values differ by 1, enter 13 in the TblStart location and 1 in the ΔTbl location. Then choose AUTO in the Indpnt: location by having the cursor on that word and pressing [ENTER].

```
TABLE SETUP
  TblStart=13
  ΔTbl=1
Indpnt: Auto Ask

[TABLE]
```

> Press [F1] [TABLE], and observe the list of input and output values. Notice that you can scroll through the table with [▼], [▲], [◄], and/or [►].

```
  x      y1
 13    11.50828
 14    12.57855
 15    13.74836
 16    15.02696
 17    16.42446
 18    17.95194
y1=13.748359355277
[TBLST][SELCT][ x ][ y ]
```

- The table values may be rounded in the table display. You can see more of the output by highlighting a particular value and viewing more decimal places at the bottom of the screen.

Return to the table set-up screen with [F1] [TBLST]. To compute specific outputs rather than a list of values, choose ASK in the Indpnt: location. Press [ENTER]. (Note that when using ASK, the settings for TblStart and ΔTbl do not matter.)	TABLE SETUP TblStart=13 ΔTbl=1 Indpnt: Auto **ASK** **TABLE**
Press [F1] [TABLE], type in the x-value(s) at which the function is to be evaluated, and press [ENTER]. Unwanted entries or values from a previous problem can be cleared with [DEL].	x | y1 15 | 13.748359 0 | 3.622 20 | 21.44627 y1=13.748359355277 **TBLST|SELCT| x | y**

NOTE: If you are interested in evaluating a function at inputs that are not evenly spaced and/or you only need a few outputs, you should use the ASK feature of the table instead of AUTO.

1.1.1g FINDING INPUT VALUES USING THE SOLVER

Your calculator solves for the input values of any equation that is in the form *"expression = constant"*. This means that all terms involving the variable must be on one side of the equation and constant terms must be on the other side before you enter the equation into the calculator. The expression can, but does not have to, use x as the input variable.

The TI-86 offers several methods of solving for input variables. We first illustrate using the SOLVER. (Solving using graphical methods will be discussed after using the SOLVER is explored.) You can refer to an equation that you have already entered in the y(x)= list or you can enter the equation in the solver.

Return to the home screen with [2nd] [EXIT] (QUIT). Access the solver by pressing [2nd] [GRAPH] (SOLVER). If there are no equations stored in the solver, you will see the screen displayed on the right – or	eqn:■ **Re3E4| y1**
– if the solver has been previously used, you will see a screen similar to the one shown on the right. If this is the case, press [▲] until only the eqn: line is on the screen and [CLEAR] to delete the old equation. You should then have the screen that is shown in the previous step.	exp=2x²–3 exp=0 x=2 bound={-1ᴇ99,1ᴇ99} **GRAPH|WIND|ZOOM|TRACE|SOLVE**

Let's now use the solver to answer the question in part *e* of Example 3 in Section 1.1: "When did the land value reach \$20,000?" Because the land value is given by $v(t) = 3.622(1.093^t)$ thousand dollars where t is the number of years after the end of 1985, we are asked to *solve* the equation $3.622(1.093^t) = 20$. That is, we are asked to find the input value t that makes this equation a true statement.

If you already have y1 = 3.622(1.093^x) in the graphing list, you can refer to the function as y1 in the SOLVER. (Note that y1 can be entered by pressing the F-key under its location in the menu at the bottom of the screen.) If not, enter 3.622(1.093^x) instead of y1 in the eqn: location. Press [ENTER]. Enter 20 in the exp: location under y1 to tell the TI-86 the rest of the equation.	exp=y1 exp=20 x=2 bound={-1ᴇ99,1ᴇ99} **GRAPH|WIND|ZOOM|TRACE|SOLVE**

If you need to edit the equation, press △ until the previous screen reappears. Edit the equation and then return here.

With the cursor in the x location, enter a guess – say 19. (You could have also used as a guess* the value that was in the x location when you accessed the SOLVER.)

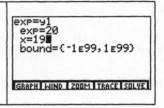

*More information on entering a guess appears at the end of this discussion.

CAUTION: You should not change anything in the "bound" location of the SOLVER. The values in that location are the ones between which the TI-86 searches[1] for a solution. If you should accidentally change anything in this location, exit the solver, and begin the entire process again. (The bound is automatically reset when you exit the SOLVER.)

Be certain that the cursor is on the line corresponding to the input variable for which you are solving (in this example, x).

Solve for the input by pressing F5 [SOLVE].

The answer to the original question is that the land value was $20,000 about 19.2 years after 1985 – *i.e.*, in the year 2005.

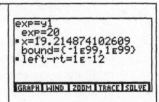

- Notice the black dot that appears next to x and next to the last line on the above screen. This is the TI-86's way of telling you that a solution has been found. When the bottom line on the screen that states left – rt ≈ 0, the value found for x is an exact solution.

- If a solution continues beyond the edge of the calculator screen, you see "…" to the right of the value. Be certain that you press and hold ▶ to scroll to the end of the number.

  The value may be given in scientific notation, and the portion that you cannot see determines the location of the decimal point. (See *Basic Operation, #3*, in this *Guide*.)

1.1.1h HOW TO DETERMINE A GUESS TO USE IN THE EQUATION SOLVER What you use in the solver as a guess tells the TI-86 where to start looking for the answer. How close your guess is to the actual answer is not very important unless there is more than one solution to the equation. *If the equation has more than one answer, the solver will return the solution that is closest to the guess you supply.* In such cases, you need to know how many answers you should search for and their approximate locations.

Three of the methods that you can use to estimate the value of a guess for an answer from the SOLVER follow. We illustrate these methods using the land value function from Example 3 of Section 1.1 and the equation $v(t) = 3.622(1.093^t) = 20$.

1. Enter the function in some location of the graphing list – say y1 = 3.622(1.093^ x) and draw a graph of the function. Press F4 [TRACE] and hold down either ▶ or ◀ until you have an estimate of where the output is 20. Use this *x*-value, 19 or 19.3 or 19.33, as your guess in the SOLVER.

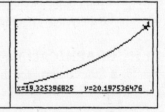

Remember that you can use any letter to represent the input in the solver, but the TI-86 will only graph a function when you use x as the input variable.

----

[1] It is possible to change the bound if the calculator has trouble finding a solution to a particular equation. This, however, should happen rarely. Refer to the *TI-86 Graphing Calculator Guidebook* for details.

2. Enter the left- and right-hand sides of the equation in two different locations of the y(x)= list – say y1 = 3.622(1.093^ x) and y2 = 20. With the graph on the screen, press F4 [TRACE] and hold down either ▶ or ◀ until you get an estimate of the X-value where the curve crosses the horizontal line representing 20.	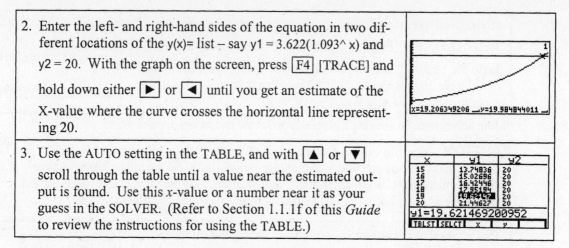
3. Use the AUTO setting in the TABLE, and with ▲ or ▼ scroll through the table until a value near the estimated output is found. Use this x-value or a number near it as your guess in the SOLVER. (Refer to Section 1.1.1f of this *Guide* to review the instructions for using the TABLE.)	

The TI-86 lets you draw a graph from which to estimate a guess for the SOLVER from within the SOLVER. When you use this feature, you are using a combination of all three of the above methods because the calculator draws a graph of *left side of equation – right side of equation* at each x-value between xMin and xMax and displays the value of the difference of the two sides of the equation as you trace the graph.

Enter the SOLVER with 2nd GRAPH (SOLVER) and press F2 [WIND]. Set window values similar to those shown to the right. (Note that some, but not all, the ZOOM options are available on the menu in the SOLVER.)	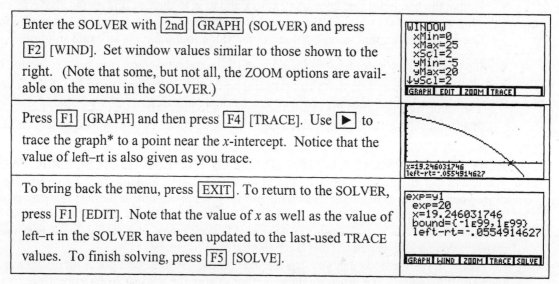
Press F1 [GRAPH] and then press F4 [TRACE]. Use ▶ to trace the graph* to a point near the x-intercept. Notice that the value of left–rt is also given as you trace.	
To bring back the menu, press EXIT. To return to the SOLVER, press F1 [EDIT]. Note that the value of x as well as the value of left–rt in the SOLVER have been updated to the last-used TRACE values. To finish solving, press F5 [SOLVE].	

*It is not easy to set a good window for this graph. However, you can still trace the graph, even if you do not see any of it on the screen. If this is the case, carefully check the left – right values to see in which direction to trace. Remember that you are looking for an input where left – right is approximately zero.

1.1.1i GRAPHICALLY FINDING INTERCEPTS Finding the input value at which the graph of a function crosses the vertical and/or horizontal axis can be found graphically or by using the SOLVER. Remember the process by which we find intercepts:

- To find the y-intercept of a function $y = f(x)$, set $x = 0$ and solve the resulting equation.
- To find the x-intercept of a function $y = f(x)$, set $y = 0$ and solve the resulting equation.

An *intercept* is the where the graph crosses or touches an axis. Also remember that the x-intercept of the function $y = f(x)$ has the same value as the root or solution of the equation $f(x) = 0$. **Thus, finding the x-intercept of the graph of $f(x) - c = 0$ is the same as solving the equation $f(x) = c$.**

We illustrate this method with a problem similar to the one in Activity 36 in Section 1.1 of *Calculus Concepts*. Suppose we are asked to find the input value of $f(x) = 3x - 0.8x^2 + 4$ that corresponds to the output $f(x) = 2.3$. That is, we are asked to find x such that $3x - 0.8x^2 + 4 = 2.3$. Because this function is not given in a context, we have no indication of an interval of input values to use when drawing the graph. So, we use the ZOOM features to set an initial view and then manually set the WINDOW until we see a graph that shows the important points of the function (in this case, the intercept or intercepts.) You can solve this equation graphically using either the *x-intercept method* or the *intersection method*. We present both, and you should use the one you prefer.

X-INTERCEPT METHOD for solving the equation $f(x) - c = 0$:

Press [GRAPH] [F1] [y(x)=] and clear all locations with [CLEAR]. Enter the function $3x - 0.8x^2 + 4 - 2.3$ in y1. You can enter x^2 with [x-VAR] [x^2] or enter it with [x-VAR] [^] 2. Remember to use [−], not [(−)], for the subtraction signs.	Plot1 Plot2 Plot3 \y1■3 x−.8 x²+4−2.3 \y2= y(x)= WIND ZOOM TRACE GRAPH x y INSf DELf SELCT►

NOTE: Whenever there are two menus at the bottom of the display screen, press [EXIT] to delete the bottom menu or press [2nd] before pressing the F-key you want to access a certain command in the top menu. We give instructions assuming there is only one menu.

Draw the graph with [F3] [ZOOM] [MORE] [F4] [ZDECM] or [F4] [ZSTD]. If you use the former, press [F2] [WIND] and reset yMax to 5 to get a better view of the graph. (If you reset the window, press [F5] [GRAPH] to draw the graph.)	y(x)= WIND ZOOM TRACE GRAPH►
To graphically find an *x*-intercept, *i.e.*, a value of *x* at which the graph crosses the horizontal axis, press [MORE] [F1] [MATH] [F1] [ROOT]. Press and hold [◄] until you are near, but to the *left* of, the leftmost *x*-intercept. Press [ENTER] to mark the location of the *left* bound for the *x*-intercept.	Left Bound? X=−.7 y=−.792
Notice the small arrowhead (▶) that appears above the location to mark the left bound. Now press and hold [►] until you are to the *right* of this *x*-intercept. Press [ENTER] to mark the location of the *right* bound for the *x*-intercept.	Right Bound? X=−.3 y=.728
For your "guess", press [◄] to move the cursor near to where the graph crosses the horizontal axis. Press [ENTER].	Guess? X=−.5 y=0
The input of the leftmost *x*-intercept is displayed as x = ⁻0.5. Note that if this process does not return the correct value for the intercept you are trying to find, you have probably not included the place where the graph crosses the axis between the two bounds (*i.e.,* between the ▶ and ◀ marks on the graph.)	ROOT X=−.5 y=0

| Repeat the above procedure to find the other x-intercept. Confirm that it occurs where x = 4.25. | 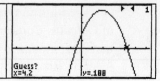 |

INTERSECTION METHOD for solving the equation $f(x) = c$:

Press GRAPH F1 [y(x)=] and clear all locations with CLEAR. Enter one side of the equation, $3x - 0.8x^2 + 4$ in y1 and the other side of the equation 2.3, in y2.	Plot1 Plot2 Plot3 \y1■3 x-.8 x²+4 \y2■2.3 y(x)= WIND ZOOM TRACE GRAPH x y INSE DELE SELCT▸
Draw the graphs with F3 [ZOOM] MORE F4 [ZDECM] or F4 [ZSTD]. If you use the former, press F2 [WIND] and reset yMax to 8 to get a better view of the graph. (If you reset the window, press F5 [GRAPH] to draw the graph.)	
To graphically find where y1 = y2, press MORE F1 [MATH] MORE F3 [ISECT]. Note that the number corresponding to the function's location in the y(x)= list is shown at the top right of the screen. Press ENTER to mark the first curve.	First curve? X=1.2 y=6.448
The cursor jumps to the other function – here, the line. Note that the number corresponding to the function's location in the y(x)= list is shown at the top right of the screen. Next, press ENTER to mark the second curve.	Second curve? X=1.2 y=2.3
Next, supply a guess for the point of intersection. Use ◀ to move the cursor near the intersection point you want to find – in this case, the leftmost point. Press ENTER.	Guess? X=-.4 y=2.3
The value of the leftmost x-intercept has the x-coordinate x = ‾0.5. Repeat the above procedure to find the rightmost x-intercept. Confirm that it is where x = 4.25.	Intersection X=‾.5 y=2.3	Intersection X=4.25 y=2.3

- Practice using each of the above methods by solving the equation $3.622(1.093^x) = 20$. Obtain further practice by solving the equation given above using the SOLVER.

1.1.1j SUMMARY OF ESTIMATING AND SOLVING METHODS   Use the method you prefer.

When you are asked to *estimate* or *approximate* an output or an input value, you can:
- Trace a graph entered in the y(x)= list   (Section 1.1.1d)
- Trace a graph showing values of the left – right sides of an equation that is entered in the SOLVER   (Section 1.1.1h)
- Use close values obtained from the TABLE   (Section 1.1.1f)

When you are asked to *find* or *determine* an output or an input value, you should:
- Evaluate an output on the graphics screen   (Section 1.1.1c)

- Evaluate an output on the home screen   (Section 1.1.1e)
- Evaluate an output value using the table   (Section 1.1.1f)
- Find an input using the solver   (Sections 1.1.1g, 1.1.1h)
- Find an input value from the graphics screen (using the *x*-intercept
  method or the intersection method)   (Section 1.1.1i)

1.2 Constructed Functions

Your calculator can find output values of and graph combinations of functions in the same way that you do these things for a single function. The only additional information you need is how to enter constructed functions in the graphing list.

1.2.1a FINDING THE SUM, DIFFERENCE, PRODUCT, QUOTIENT, OR COMPOSITE FUNCTION
Suppose that a function *f* has been entered in y1 and a function *g* has been entered in y2. Your calculator will evaluate and graph these constructed functions:

Enter y1 + y2 in y3 to obtain the sum function $(f + g)(x) = f(x) + g(x)$.

Enter y1 – y2 in y4 to obtain the difference function $(f - g)(x) = f(x) - g(x)$.

Enter y1*y2 in y5 to obtain the product function $(f \cdot g)(x) = f(x) \cdot g(x)$.

Enter y1/y2 in y6 to obtain the quotient function $(f \div g)(x) = \dfrac{f(x)}{g(x)}$.

Enter y1(y2) in y7 to obtain the composite function $(f \circ g)(x) = f(g(x))$.

The functions can be entered in any location in the y(x)= list. Although the TI-86 will not give an algebraic formula for a constructed function, you can check your final answer by evaluating your constructed function and the calculator-constructed function at several different points to see if they yield the same outputs.

1.2.1b FINDING A DIFFERENCE FUNCTION
We illustrate this technique with the functions that are given on page 19 of Section 1.2 of *Calculus Concepts*: Sales = $S(t) = 3.570(1.105^t)$ million dollars and costs = $C(t) = -39.2t^2 + 540.1t + 1061.0$ thousand dollars *t* years after 1996.

Press GRAPH F1 [y(x)=], and clear each previously-entered equation with CLEAR or F4 [DELf]. Enter *S* in y1 with 3 ⎵. 570 (1 ⎵. 105 ^ x-VAR) ENTER . Enter *C* in y2 with (–) 39 ⎵. 2 x-VAR x^2 + 540 ⎵. 1 x-VAR + 1061 ENTER .
The difference function, the profit $P(x) = S(x) - 0.001C(x) =$ y1 – 0.001 y2, is entered in y3 with 2nd ALPHA (alpha) 0 (y) 1 ⎵ ⎵. 001 2nd ALPHA (alpha) 0 (y) 2.

- Note that you can also type y when entering the third function in the graphing list by pressing the F-key corresponding to its position on the menu bar.

To find the profit in 1998, evaluate y3 when x = 2. You can evaluate on the home screen, the graphics screen, or in the table. We choose to use the home screen.

Return to the home screen with [2nd] [EXIT] (QUIT). Press
[2nd] [ALPHA] (alpha) 0 (y) 3 [(] 2 [)] [ENTER] to see the result.
We find that the profit in 1998 was $P(2) \approx \$2.375$ million.

```
y3(2)
            2.37465925
```

1.2.1c FINDING A PRODUCT FUNCTION   We illustrate this technique with the functions that given on page 21 of Section 1.2 of *Calculus Concepts*: Milk price = $M(x) = 0.007x + 1.492$ dollars per gallon on the xth day of last month and milk sales = $G(x) = 31 - 6.332(0.921^x)$ gallons of milk sold on the xth day of last month.

Press [GRAPH] [F1] [y(x)=], and clear each previously entered equation with [CLEAR] or [F4] [DELf]. Enter M in y1 using [.] 007 [x-VAR] [+] 1 [.] 492 [ENTER] and enter G in y2 with 31 [−] 6 [.] 332 [(] [.] 921 [^] [x-VAR] [)] [ENTER].

```
Plot1 Plot2 Plot3
\y1∎.007 x+1.492
\y2∎31-6.332(.921^x)
\y3=∎

y(x)= WIND  ZOOM TRACE GRAPH
 x   y    INSf DELf SELCT
```

Enter the product function $T(x) = M(x) \cdot G(x) = $ y1·y2 in y3 with the keystrokes [2nd] [ALPHA] (alpha) 0 (y) 1 [×] [2nd] [ALPHA] (alpha) 0 (y) 2.

```
Plot1 Plot2 Plot3
\y1∎.007 x+1.492
\y2∎31-6.332(.921^x)
\y3∎y1*y2

y(x)= WIND  ZOOM TRACE GRAPH
 x   y    INSf DELf SELCT
```

NOTE: You do not have to use [×] between y1 and y2 to indicate a product function. You cannot use parentheses to indicate the product function because the TI-86 will think that you are entering y3 as a composite function.

To find milk sales on the 5th day of last month, evaluate y3 at $x = 5$. We choose to do this on the home screen. Return to the home screen with [2nd] [MODE] (QUIT). Press [2nd] [ALPHA] (alpha) 0 (y) 3 [(] 5 [)] [ENTER] to see the result. We find that milk sales were $T(5) \approx \$40.93$.

```
y3(5)
          40.9296552175
```

1.2.2a CHECKING YOUR ANSWER FOR A COMPOSITE FUNCTION   We illustrate this technique with the functions that are given on page 23 of Section 1.2 of *Calculus Concepts*: altitude = $F(t) = -222.22t^3 + 1755.95t^2 + 1680.56t + 4416.67$ feet above sea level where t is the time into flight in minutes and air temperature = $A(F) = 277.897(0.99984^F) - 66$ degrees Fahrenheit where F is the number of feet above sea level. Remember that when you enter functions in the y(x)= list, you must use x as the input variable.

Clear the functions from the y(x)= list. Enter F in y1 by pressing [(−)] 222 [.] 22 [x-VAR] [^] 3 [+] 1755 [.] 95 [x-VAR] [x^2] [+] 1680 [.] 56 [x-VAR] [+] 4416 [.] 67 [ENTER]. Enter A in y2 by pressing 277 [.] 897 [(] [.] 99984 [^] [x-VAR] [)] − 66 [ENTER].

```
Plot1 Plot2 Plot3
\y1∎-222.22x^3+1755....
\y2∎277.897(.99984^x...
\y3=∎

y(x)= WIND  ZOOM TRACE GRAPH
 x   y    INSf DELf SELCT
```

Enter the composite function $(A \circ F)(x) = A(F(x)) = $ y2(y1) in y3 with [2nd] [ALPHA] (alpha) 0 (y) 2 [(] [2nd] [ALPHA] (alpha) 0 (y) 1 [)] [ENTER].

```
Plot1 Plot2 Plot3
\y1∎-222.22x^3+1755....
\y2∎277.897(.99984^x...
\y3∎y2(y1)
\y4=∎

y(x)= WIND  ZOOM TRACE GRAPH
 x   y    INSf DELf SELCT
```

Next, enter your algebraic answer for the composite function in y4. Be certain that you enclose the exponent in y4 (the function in y1) in parentheses!

(The composite function in the text is the one that appears to the right, but you should enter the function that you have found for the composite function.)

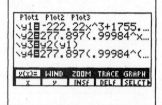

We now wish to check that the algebraic answer for the composite function is the same as the calculator's composite function by evaluating both functions at several different input values. You can do these evaluations on the home screen, but as seen below, using the table is more convenient.

1.2.2b **TURNING FUNCTIONS OFF AND ON IN THE GRAPHING LIST** Note in the prior illustration that we are interested in the output values for only y3 and y4. However, the table will list values for all functions that are turned on. (A function is *turned on* when the equals sign in its location in the graphing list is darkened.) We now wish to turn off y1 and y2.

Press $\boxed{\text{GRAPH}}$ $\boxed{\text{F1}}$ [y(x)=] and place the cursor in the line containing y1. Press $\boxed{\text{F5}}$ [SELCT] and use $\boxed{\blacktriangledown}$ to move the cursor to the y2 line. Press $\boxed{\text{F5}}$ [SELCT]; y1 and y2 are now turned off.

A function is *turned off* when the equals sign in its location in the graphing list is not dark. To turn a function back on, simply reverse the above process to make the equal sign for the function dark. The SELCT key toggles between the function being off and being on. When you draw a graph, the TI-86 graphs of all functions that are turned on. You may at times wish to keep certain functions entered in the graphing list but not have them graph and not have their values shown in the table. Such is the case in this illustration. We now return to checking to see that y3 and y4 represent the same function.

Choose the ASK setting in the table setup so that you can check several different values for both y3 and y4. Recall that you access the table setup with $\boxed{\text{TABLE}}$ $\boxed{\text{F2}}$ [TBLST]. Move the cursor to ASK in the Indpnt: location and press $\boxed{\text{ENTER}}$.

TABLE SETUP
TblStart=15
ΔTbl=1
Indpnt: Auto **ASK**
TABLE

Press $\boxed{\text{F1}}$ (TABLE), type in the *x*-value(s) at which the function is to be evaluated, and press $\boxed{\text{ENTER}}$ after each one. We see that because all these outputs are the same for each function, you can be fairly sure that your answer is correct.

x	y3	y4
-5	-65.9945	-65.9945
0	71.07481	71.07481
8	-46.0627	-46.0627
20	3.406E74	3.406E74
57	ERROR	ERROR
y3=3.4061205793316E74		
TBLST **SELCT** x y		

Why does ERROR appear in the table when $x = 57$? Look at the value when $x = 20$; it is very large! The computational limits of the calculator have been exceeded when $x = 57$. The TI-86 calls this an OVERFLOW error.

1.2.3 **GRAPHING A PIECEWISE CONTINUOUS FUNCTION** Piecewise continuous functions are used throughout the text. You will need to use your calculator to graph and evaluate outputs of piecewise continuous functions. Several methods can be used to draw the graph of a piecewise function. One of these is presented below using the function that appears in Example 2 of Section 1.2 in *Calculus Concepts*:

The population of West Virginia from 1985 through 1999 can be modeled by

$$P(t) = \begin{cases} -23.373t + 3892.220 \text{ thousand people} & \text{when } 85 \le t < 90 \\ -1.013t^2 + 193.164t - 7387.836 \text{ thousand people} & \text{when } 90 \le t \le 99 \end{cases}$$

where t is the number of years since 1900.

Clear any functions that are in the y(x)= list. Using x as the input variable, enter each piece of the function in a separate location. We use locations y1 and y2. Next, we form the formula for the piecewise function in y3.	

Parentheses <u>must</u> be used around the function portions and the inequality statements that tell the calculator which side of the break point to graph each part of the piecewise function.

Have the cursor in y2 and press ▼ to place the cursor in y3. Press ([ ] [2nd] [ALPHA] (alpha) 0 (Y) 1)] [( ] [x-VAR] [2nd] 2 (TEST) [F2] [<] 90)] [+] [( ] [2nd] [ALPHA] (alpha) 0 (Y) 2)] [( ] [x-VAR] [2nd] 2 (TEST) [F5] [≥] 90)] [ENTER].	

Your calculator draws graphs by connecting function outputs wherever the function is defined. However, this function breaks at $x = 90$. The TI-83 will connect the two pieces of P unless you tell it not to do so. Whenever you draw graphs of piecewise functions, set your calculator to Dot mode as described below so that it will not connect the function pieces at the break point.

Turn off y1 and y2 and place the cursor on y3. Press [MORE] [F3] [STYLE]. Press [F3] five more times and the slanted line[2] to the *left* of y3 should be a dotted line (as shown to the right). The function y3 is now in Dot mode.	

NOTE: The method described above places individual functions in Dot mode. The functions return to standard (Connected) mode when the function locations are cleared. If you want to put all functions in Dot mode at the same time, press [GRAPH] [F2] [WIND] [MORE] [F3] [FORMT], choose DrawDot in the third line, and press [ENTER]. However, if you choose to set DOT mode in this manner, you must return to the window format screen, select DrawLine instead of DrawDot, and press [ENTER] to take the TI-86 out of Dot mode.

Now, set the window. The function P is defined only when the input is between 85 and 99. So, we evaluate $P(85)$, $P(90)$, and $P(99)$ to help when setting the vertical view.	y3(85) 1905.515
	 y3(90)
	               1791.624
	 y3(99)
	               1806.987

Note that if you attempt to set the window using ZFIT as described in Section 1.1. 1b of this *Guide*, the picture is not very good and you will probably want to manually reset the height of the window as described below.

---

[2] The different graph styles you can draw from this location are described in more detail on page 79 in your *TI-86 Graphing Calculator Guidebook*.

We set the lower and upper endpoints of the input interval as xMin and xMax, respectively. Press $\boxed{\text{GRAPH}}$ $\boxed{\text{F2}}$ [WIND], set xMin = 85, xMax = 99, yMin ≈ 1780, and yMax ≈ 1910. Press $\boxed{\text{F5}}$ [GRAPH] and use $\boxed{\text{CLEAR}}$ to remove the menu.	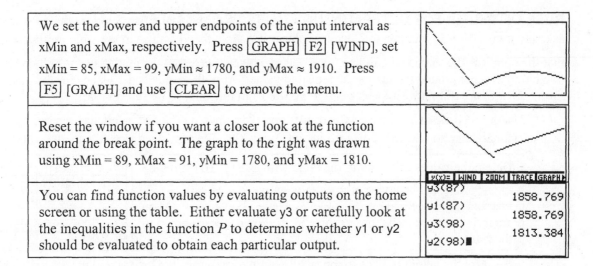
Reset the window if you want a closer look at the function around the break point. The graph to the right was drawn using xMin = 89, xMax = 91, yMin = 1780, and yMax = 1810.	
You can find function values by evaluating outputs on the home screen or using the table. Either evaluate y3 or carefully look at the inequalities in the function P to determine whether y1 or y2 should be evaluated to obtain each particular output.	y3(87) 1858.769 y1(87) 1858.769 y3(98) 1813.384 y2(98)■

1.3 Limits: Functions, Limits, and Continuity

The TI-86 table is an essential tool when you estimate end behavior numerically. Even though rounded values are shown in the table due to space limitations, the TI-86 displays at the bottom of the screen many more decimal places for a particular output when you highlight that output.

1.3.1a NUMERICALLY ESTIMATING END BEHAVIOR   Whenever you use the TI-86 to estimate end behavior, set the TABLE to ASK mode. We illustrate using the function u that appears in Example 1 of Section 1.3 in *Calculus Concepts*:

Press $\boxed{\text{GRAPH}}$ $\boxed{\text{F1}}$ [y(x)=] and use $\boxed{\text{F4}}$ [DELf] to delete all previously-entered functions. Enter $u(x) = \dfrac{3x^2 + x}{10x^2 + 3x + 2}$. Be certain to enclose both numerator and denominator of the fraction in parentheses.	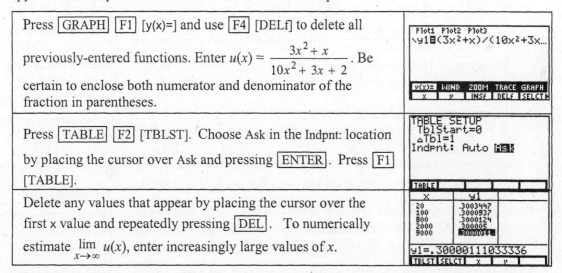
Press $\boxed{\text{TABLE}}$ $\boxed{\text{F2}}$ [TBLST]. Choose Ask in the Indpnt: location by placing the cursor over Ask and pressing $\boxed{\text{ENTER}}$. Press $\boxed{\text{F1}}$ [TABLE].	
Delete any values that appear by placing the cursor over the first x value and repeatedly pressing $\boxed{\text{DEL}}$. To numerically estimate $\lim\limits_{x \to \infty} u(x)$, enter increasingly large values of x.	

NOTE: The values you enter do not have to be those shown in the text or these shown in the above table provided the values you input increase without bound.

CAUTION: Your instructor will very likely have you write the table you construct on paper. Be certain that if necessary, you highlight the rounded values in the output column of the table and look on the bottom of the screen to see what these values actually are.

ROUNDING OFF: Recall that *rounded off* (also called *rounded* in this *Guide*) means that if one digit past the digit of interest if less than 5, other digits past the digit of interest are dropped. If one digit past the one of interest is 5 or over, the digit of interest is increased by 1 and the remaining digits are dropped.

RULE OF THUMB FOR DETERMINING LIMITS FROM TABLES: Suppose that you are asked to give $\lim\limits_{x \to \infty} u(x)$ accurate to 3 decimal places. Observe the y1 values in the table until you see that the output is the same value to one more decimal place (here, to 4 decimal places) for 3 consecutive outputs. Then, round the last of the repeated values off to the requested 3 places for the desired limit. Your instructor may establish a different rule from this one, so be sure to ask.

Using this Rule of Thumb and the results that are shown on the last calculator screen, we estimate that, to 3 decimal places, $\lim\limits_{x \to \infty} u(x) = 0.300$. We now need to estimate $\lim\limits_{x \to {}^-\infty} u(x)$.

Delete the values currently in the table with DEL . To estimate the negative end behavior of u, enter negative values with increasingly large magnitudes. (*Note*: Again, the values that you enter do not have to be those shown in the text or these shown to the right.)	

Because the output 0.2999... appears three times in a row, we estimate $\lim\limits_{x \to {}^-\infty} u(x) = 0.300$.

CAUTION: It is not the final value, but a sequence of several values, that is important when determining limits. If you enter a very large or very small value, you may exceed the limits of the TI-86's capability and obtain an incorrect number. Always look at the sequence of values obtained to make sure that all values that are found make sense.

1.3.1b GRAPHICALLY ESTIMATING END BEHAVIOR The graph of the function u can be used to confirm our numerically estimated end behavior. Even though the ZOOM menu of the TI-86 can be used with this process for some functions, the graph of u is lost if you use ZOUT in the ZOOM menu. We therefore manually set the window to zoom out on the horizontal axis.

Have $u(x) = \dfrac{3x^2 + x}{10x^2 + 3x + 2}$ in the y1 location of the y(x)= list. (Be certain that you remember to enclose both the numerator and denominator of the fraction in parentheses.) A graph drawn with GRAPH F3 [ZOOM] MORE F4 [ZDECM] is a starting point.	
We estimated the limit as x gets very large or very small to be 0.3. Now, $u(0) = 0$, and it does appears from the graph that u is never negative. Set a window with values such as xMin = ⁻10, xMax = 10, yMin = 0, and yMax = 0.35. Press F5 [GRAPH].	
To examine the limit as x gets larger and larger (*i.e.,* to *zoom out* on the positive x-axis), change the window so that xMax = 100, view the graph with F5 [GRAPH], change the window so that xMax = 1000, view the graph with F5 [GRAPH], and so forth. Press F4 [TRACE] and hold down ▶ on each graph screen to view the outputs.	 The sequence of observed output values confirm our numerical estimate.
Repeat the process as x gets smaller and smaller, but change xMin rather than xMax after drawing each graph. The graph to the right was drawn with xMin = ⁻10,000, xMax = 10, yMin = 0, and yMax = 0.35. Press F4 [TRACE] and hold ◀ while on each graph screen to view some of the outputs and confirm the numerical estimates.	

1.3.2a NUMERICALLY ESTIMATING THE LIMIT AT A POINT Whenever you numerically estimate the limit at a point, you should again set the TABLE to ASK mode. We illustrate using the function *u* that appears in Example 2 of Section 1.3 in *Calculus Concepts*:

Have $u(x) = \dfrac{3x}{9x + 2}$ in some location of the y(x)= list, say y1.
 	
Have TBLST set to Ask, and press ⌑TABLE⌑ ⌑F1⌑ [TABLE] to return to the table.
	```Plot1 Plot2 Plot3``` `\y1▄3x/(9x+2)`
	
	 `y(x)= WIND ZOOM TRACE GRAPH`
	 `x y INSf DELf SELCT▸`

Delete the values currently in the table with ⌑DEL⌑. To numerically estimate $\displaystyle\lim_{x \to -2/9^-} u(x)$, enter values to the *left* of, and *closer and closer* to, $-2/9 = -0.222222...$. Because the output values appear to become larger and larger, we estimate that the limit does not exist and write $\displaystyle\lim_{x \to -2/9^-} u(x) \to \infty$.	```x y1```
	 `-.23     9.857143`
	 `-.223    95.57143`
	 `-.2223   952.7143`
	 `-.22223  8524.143`
	 `-.222223 95238.428`
	 `y1=95238.428571429`
	 `TBLST SELCT x y`

Delete the values currently in the table. To numerically estimate $\displaystyle\lim_{x \to -2/9^+} u(x)$, enter values to the *right* of, and *closer and closer* to, $-2/9$. Because the output values appear to become larger and larger, we estimate that $\displaystyle\lim_{x \to -2/9^+} u(x) \to -\infty$.	```x y1```
	 `-.21     -5.72727`
	 `-.218    -17.2105`
	 `-.2218   -175.105`
	 `-.22218  -1754.05`
	 `-.222218 -17543.5`
	 `y1=-17543.526315789`
	 `TBLST SELCT x y`

1.3.2b GRAPHICALLY ESTIMATING THE LIMIT AT A POINT A graph can be used to estimate a limit at a point or to confirm a limit that you estimate numerically. The procedure usually involves *zooming in* on a graph to confirm that the limit at a point exists or *zooming out* to validate that a limit does not exist. We again illustrate using the function *u* that appears in Example 2 of Section 1.3 in *Calculus Concepts*.

Have the function $u(x) = \dfrac{3x}{9x + 2}$ entered in some location of the y(x)= list, say y1. A graph drawn with ⌑GRAPH⌑ ⌑F3⌑ [ZOOM] ⌑F4⌑ [ZSTD] or with ⌑GRAPH⌑ ⌑F3⌑ [ZOOM] ⌑MORE⌑ ⌑F4⌑ [ZDECM] is not very helpful.	(graph)

To confirm that $\displaystyle\lim_{x \to -2/9^-} u(x)$ and $\displaystyle\lim_{x \to -2/9^+} u(x)$ do not exist, we are interested in values of *x* that are near to and on either side of $-2/9$.

Choose input values close to $-0.222222...$ for the *x*-view and experiment with different *y* values until you find an appropriate vertical view. Use these values to manually set a window such as that shown to the right. Draw the graph with ⌑F5⌑ [GRAPH].	```WINDOW```
	 `xMin=-.42`
	 `xMax=-.02`
	 `xScl=1`
	 `yMin=-7`
	 `yMax=7`
	 `↓yScl=1`
	 `y(x)= WIND ZOOM TRACE GRAPH▸`

The vertical line appears because the TI-86 is set to Connected mode. Place the TI-86 in DrawDot mode or place the function y1 in the y(x)= list in Dot mode (see page B16) and redraw the graph.	(two graphs)

It appears from this graph that as x approaches $-2/9$ from the left that the output values increase without bound and that as x approaches $-2/9$ from the right that the output values decrease without bound. Choosing smaller yMin values and larger yMax values in the Window and tracing the graph as x approaches $-2/9$ from either side confirms this result.

Graphically Estimating a Limit at a Point when the Limit Exists: The previous illustrations involved zooming on a graph by manually setting the window. You can also zoom with the ZOOM menu of the calculator. We next describe this method by *zooming in* on a function for which the limit at a point exists.

Have the function $h(x) = \dfrac{3x^2 + 3x}{9x^2 + 11x + 2}$ entered in the y1 location of the y(x)= list. Suppose that we want to estimate $\lim\limits_{x \to -1} h(x)$.

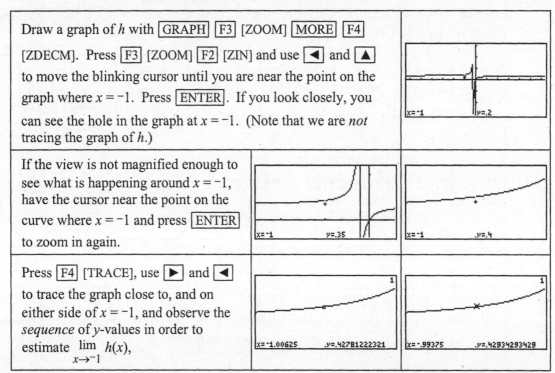

Draw a graph of h with [GRAPH] [F3] [ZOOM] [MORE] [F4] [ZDECM]. Press [F3] [ZOOM] [F2] [ZIN] and use [◄] and [▲] to move the blinking cursor until you are near the point on the graph where $x = -1$. Press [ENTER]. If you look closely, you can see the hole in the graph at $x = -1$. (Note that we are *not* tracing the graph of h.)

If the view is not magnified enough to see what is happening around $x = -1$, have the cursor near the point on the curve where $x = -1$ and press [ENTER] to zoom in again.

Press [F4] [TRACE], use [►] and [◄] to trace the graph close to, and on either side of $x = -1$, and observe the *sequence* of y-values in order to estimate $\lim\limits_{x \to -1} h(x)$,

Observing the sequence of y-values is the same procedure as numerically estimating the limit at a point. Therefore, it is not the value at $x = -1$ that is important; the limit is what the output values displayed on the screen approach as x approaches -1. It appears that $\lim\limits_{x \to -1^-} h(x) \approx 0.43$ and $\lim\limits_{x \to -1^+} h(x) \approx 0.43$. Therefore, we conclude that $\lim\limits_{x \to -1} h(x) \approx 0.43$.

1.4 Linear Functions and Models

This portion of the *Guide* gives instructions for entering real-world data into the calculator and finding familiar function curves to fit that data. You will use the beginning material in this section throughout all the chapters in *Calculus Concepts*.

CAUTION: Be very careful when you enter data in your calculator because your model and all of your results depend on the values that you enter! Always check your entries.

1.4.1a ENTERING DATA We illustrate data entry using the values in Table 1.19 in Section 1.4 of *Calculus Concepts*:

Year	1999	2000	2001	2002	2003	2004
Tax (dollars)	2532	3073	3614	4155	4696	5237

Press 2nd + (STAT) F2 [EDIT] to access the lists that hold data. You see only the first 3 lists, (L1, L2, and L3) but you can access the other 2 lists (L4 and L5) by having the cursor on the list name and pressing ▶ several times. If you do not see these list names, return to the statistical setup instructions on page B-1 of this *Guide*.

In this text, we usually use list L1 for the input data and list L2 for the output data. If there are any data already in your lists, see Section 1.4.1c of this *Guide* and first delete these "old" data values. To enter data in the lists, do the following:

Position the cursor in the first location in list L1. Enter the input data into list L1 by typing the years from top to bottom in the L1 column, pressing ENTER after each entry.

After typing the sixth input value, 2004, use ▶ to cause the cursor go to the top of list L2. Enter the output data in list L2 by typing the entries from top to bottom in the L2 column, pressing ENTER after each tax value.

Data can also be entered from the home screen rather than in the statistical lists (as was shown above). The end result is the same, so you should choose the method that you prefer.

On the home screen, press 2nd – (LIST) F1 [{], and then type in each of the input data separated by commas. End the list with F2 [}]. Store this list to the name L1 with STO▶ 7 (L) ALPHA 1 ENTER. Repeat the process to enter the output data, but store these data in a list named L2.

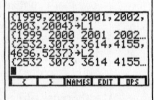

WARNING: If you do not enter and store the input data into a list named L1 and the output data into a list named L2, many of the programs in this *Guide* will not properly execute.

1.4.1b EDITING DATA No matter how it is entered, the easiest way to edit data is using the statistical lists. If you incorrectly type a data value, use the cursor keys (*i.e.,* the arrow keys ▶, ◀, ▲, and/or ▼) to darken the value you wish to correct and then type the correct value. Press ENTER.

- To *insert* a data value, put the cursor over the value that will be directly below the one you will insert, and press 2nd DEL (INS). The values in the list below the insertion point move down one location and a 0 is filled in at the insertion point. Type the data value to be inserted over the 0 and press ENTER. The 0 is replaced with the new value.

- To *delete* a single data value, move the cursor over the value you wish to delete, and press DEL. The values in the list below the deleted value move up one location.

1.4.1c DELETING OLD DATA   Whenever you enter new data in your calculator, you should first delete any previously entered data. There are several ways to do this, and the most convenient method is illustrated below.

| Access the data lists with [2nd] [+] (STAT) [F2] [EDIT]. (You probably have different values in your lists if you are deleting "old" data.) Use [▲] to move the cursor over the name L1. |
| Press [CLEAR] [ENTER]. Use [▶] and [▲] to move the cursor over the name L2. Press [CLEAR] [ENTER]. Repeat this procedure to clear the other lists. |

1.4.1d FINDING FIRST DIFFERENCES   When the input values are evenly spaced, you can use program DIFF to compute first differences in the output values. Program DIFF is given in the *TI-86 Program Appendix* at the *Calculus Concepts* web site. Consult the Programs category in *Trouble Shooting the TI-86* in this *Guide* if you have questions about obtaining the programs.

| Have the data given in Table 1.19 in Section 1.4 of *Calculus Concepts* entered in your calculator. (See Section 1.4.1a of this *Guide*.) Exit the list menu with [2nd] [EXIT] (QUIT). |
| To run the program, press [PRGM] [NAMES] followed by the F-key that is under the DIFF program location, and press [ENTER]. The message on the right appears on your screen. |
| If you have not entered the data in L1 and L2, press [F2] [Quit] and do so. Otherwise, press [F1] [Yes] to continue. Press [F1] [1st] to compute the first differences. Choose [F4] to quit. | We use the other choices in the next chapter. | You can also view the first differences in list L3. |

- The first differences are constant, so a linear function gives a perfect fit to these tax data.

You will find program DIFF very convenient to use in the next chapter because of the other options it has. However, if you do not want to make use program DIFF at this time, you can use a built-in capability of your TI-86 to compute the first differences of any list.

| Be certain that you have the output data entered in list L2 in your calculator. Press [2nd] [−] (LIST) [F5] [OPS] [MORE] [MORE] [F4] [Deltal] [ALPHA] 7 (L) 2 [)] [ENTER]. |

For larger data sets use [▶] to scroll to the right to see the remainder of the first differences. Use [◀] to scroll back to the left.

NOTE: Program DIFF **should not** be used for data with input values (entered in L1) that are *not* evenly spaced. First differences give no information about a possible linear fit to data with inputs that are not the same distance apart. If you try to use program DIFF with input data that are not evenly spaced, the message INPUT VALUES NOT EVENLY SPACED appears and the program stops. The Deltalst option discussed above gives first differences even if the input values are not evenly spaced. However, these first differences have no interpretation when finding a linear function to fit data if the input values are not evenly spaced.

1.4.2a SCATTER PLOT SETUP   The first time that you draw a graph of data, you need to set the TI-86 to draw the type of graph you want to see. Once you do this, you never need to do this set up again (unless for some reason the settings are changed). If you always put input data in list L1 and output data in list L2, you can turn the scatter plots off and on from the y(x)= screen rather than the STAT PLOTS screen after you perform this initial setup.

Press $\boxed{2nd}$ (STAT) $\boxed{F3}$ [PLOT] to display the STAT PLOTS screen. (Your screen may not look exactly like this one.) Press $\boxed{F1}$ [PLOT1].	STAT PLOTS 1:Plot1…Off ∟˙ xStat yStat 2:Plot2…Off ∟˙ xStat yStat 3:Plot3…Off ∟˙ xStat yStat `PLOT1 PLOT2 PLOT3 P1On P1Off`	On **Off** Type=∟˙ Xlist Name=xStat Ylist Name=yStat Mark=□ `PLOT1 PLOT2 PLOT3 P1On P1Off`

Press $\boxed{ENTER}$ to turn Plot1 on, and choose the options shown on the right. (You can choose any of the 3 marks at the bottom of the screen.)	Choose these options:
	$\boxed{F1}$ [SCAT], $\boxed{F4}$ [L1],
	$\boxed{F5}$ [L2], and a mark.
		On Off
		Type=∟˙
		Xlist Name=L1
		Ylist Name=L2
		Mark=□
		`PLOT1 PLOT2 PLOT3 P1On P1Off`
		`□  +  ·`

Press $\boxed{GRAPH}$ $\boxed{F1}$ [y(x)=] and notice that Plot1 at the top of the screen is darkened. This tells you that Plot1 is turned on and ready to graph data. Press $\boxed{2nd}$ $\boxed{EXIT}$ (QUIT) to return to the home screen.		**Plot1** Plot2 Plot3
		\y1=■
		
		`y(x)= WIND ZOOM TRACE GRAPH`
		`x  y  INSf DELf SELCTf`

- A scatter plot is *turned on* when its name on the y(x)= screen is darkened. To turn Plot1 off, use $\boxed{\blacktriangle}$ to move the cursor to the Plot1 location, and press $\boxed{ENTER}$. Reverse the process to turn Plot1 back on.

- You can enter the names of any data lists as the Xlist Name and the Ylist Name and draw a scatter plot. (However, it is easiest to always work with L1 and L2.) All data lists on the TI-86 can be named and stored in the calculator's memory for later recall and use. Refer to Sections 1.4.3c and 1.4.3d of this *Guide* for instructions on storing data lists and later recalling them for use.

1.4.2b DRAWING A SCATTER PLOT OF DATA   Any functions that are turned on in the y(x)= list will graph when you plot data. Therefore, you should clear, delete, or turn off all functions before drawing a scatter plot. We illustrate how to graph data using the modified tax data that follows Example 2 in Section 1.4 of *Calculus Concepts*.

Year	1999	2000	2001	2002	2003	2004
Tax (in dollars)	2541	3081	3615	4157	4703	5242

Access the y(x)= graphing list. If any entered function is no longer needed, clear it by moving the cursor to its location and pressing $\boxed{CLEAR}$. If you want the function(s) to remain but not graph when you draw the scatter plot, refer to Section 1.2.2b of this *Guide* for instructions on how to turn function(s) off. Also be sure that Plot 1 on the top left of the y(x)= screen is darkened.

Press 2nd + (STAT) F2 [EDIT]. Using the given table, enter the year data in L1 and the modified tax data in L2 according to the instructions given in Sections 1.4.1a-c of this *Guide*. (You can either leave values in the other lists or clear them.)	L1 L2 L3 2 1999 2541 -------- 2000 3081 2001 3615 2002 4157 2003 4703 2004 5242 L2(6) =5242 { } NAMES " OPS ▶
Press GRAPH F3 [ZOOM] MORE F5 [ZDATA] to have the TI-86 set an autoscaled view of the data and draw the scatter plot. Note that ZDATA does not reset the *x*- and *y*-axis tick marks. You should do this manually with the window settings if you want different spacing between the tick marks.)	

NOTE: If the menu at the bottom of the TI-86 screen obscures your view, remember that pressing CLEAR removes the menu and pressing GRAPH makes the menu reappear.

Recall that if the data are perfectly linear (that is, every data point falls on the graph of a line), the first differences in the output values are constant. The first differences for the original tax data were constant at $541, so a linear function fit the data perfectly. What information is given by the first differences for these modified tax data?

Run program DIFF by pressing PRGM F1 [NAMES] followed by F-key under DIFF, and press ENTER. Press F4 [Quit] to exit the program. View the first differences on the home screen or press 2nd + (STAT) F2 [EDIT] to view them in list L3.	Choice? 1st differences in L3 {540 534 542 546 539} 1st 2nd % Quit

These first differences are close to being constant. This information, together with the straight line pattern shown by the scatter plot, are a good indication that a linear function is likely to give a good fit to the data.

1.4.2c FINDING A LINEAR FUNCTION TO MODEL DATA Throughout this course, you will often have your TI-86 find a linear function of the form $y = a + bx$ that best fits a set of data. In this equation, a is the y-intercept and b is the slope. This equation is not of the same form, but is equivalent to, the linear equation form given in your text.

Press 2nd EXIT (QUIT) to return to the home screen. Then, press 2nd + (STAT) F1 [CALC] F3 [LinR] ALPHA 7 (L) 1 , ALPHA 7 (L) 2 , 2nd ALPHA (alpha) 0 (y) 1.	LinR L1,L2,y1 CALC EDIT PLOT DRAW VARS OneVa TwoVa LinR LnR ExpR ▶
The keystrokes above find the linear equation of best fit using L1 as the input data, L2 as the output data, and pastes the equation in y1. Press ENTER.	LinReg y=a+bx a=-1077663.6 b=540.371429 ↓corr=.999995458 CALC EDIT PLOT DRAW VARS OneVa TwoVa LinR LnR ExpR ▶
The linear equation of best fit for the modified tax data that was entered into lists L1 and L2 in Section 1.4.2b is displayed on the home screen and has been copied into the y(x)= list. (Remember that the equation scrolls to the right when you press ▶.)	Plot1 Plot2 Plot3 \y1■1077663.5809525... y(x)= WIND ZOOM TRACE GRAPH x y INSf DELf SELCT▶

CAUTION: The best-fit function found by the calculator is also called a *regression function*. The coefficients of the regression function ***never should be rounded*** when you are going to use it! This is not a problem because the calculator pastes the entire equation it finds into the graphing list at the same time the function is found if you follow the instructions given above.

NOTE: The TI-86 will use lists called xStat and yStat, which probably contain different data, if you do not specify lists L1 and L2 in the instruction to find the best-fit equation. It is possible to use lists other than L1 and L2 for the input and output data. However, if you do so, you must set one of the STAT PLOT locations to draw the scatter plot for those other lists (as described in Section 1.4.2b). To find the best-fit function, replace L1 and L2 by the other lists in the fit instruction. To paste the function into a location other than y1, just change the number 1 following y in the fit instructions to the number corresponding to the graphing location that you want.

CAUTION: The r that is shown on the screen that first gives the linear equation is called the *correlation coefficient*. This and a quantity called r^2, the *coefficient of determination*, are numbers that you will learn about in a statistics course. It is not appropriate[3] to make use of these values in a calculus course.

Graphing the Line of Best Fit: After finding a best-fit equation, you should always draw the graph of the function on a scatter plot to verify that the function provides a good fit to the data.

Press EXIT F5 [GRAPH] to overdraw the function you pasted in the graphing list on the scatter plot of the data. (As we suspected from looking at the scatter plot and the first differences, this function provides a very good fit to the data.)

1.4.2d COPYING A GRAPH TO PAPER

Your instructor may ask you to copy what is on your graphics screen to paper. If so, use the following ideas to more accurately perform this task.

After using a ruler to place and label a scale (*i.e.*, tick marks) on your paper, use the trace values (as shown below) to draw a scatter plot and graph of the line on your paper.

Press GRAPH to return the modified tax data graph found in Section 1.4.2c to the screen. Press F4 [TRACE] and ▶. The symbol P1 in the upper right-hand corner of the screen indicates that you are tracing the scatter plot of the data in Plot 1.
Press ▼ to move the trace cursor to the linear function graph. The number in the top right of the screen tells you the location of the function that you are tracing (in this case, y1). Use ▶ and/or ◀ to locate values that are as "nice" as possible and mark those points on your paper. Use a ruler to connect the points and draw the line.

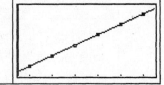

- If you are copying the graph of a continuous curve rather than a straight line, you need to trace as many points as necessary to see the shape of the curve while marking the points on your paper. Connect the points with a smooth curve.

---

[3] Unfortunately, there is no single number that can be used to tell whether one function better fits data than another. The correlation coefficient only compares linear fits and should not be used to compare the fits of different types of functions. For the statistical reasoning behind this statement, read the references in footnote 6 on page B-30.

1.4.3a ALIGNING DATA   We return to the modified tax data entered in Section 1.4.2b. If you want L1 to contain the number of years after a certain year instead of the actual year, you need to *align* the input data. In this illustration, we shift all of the data points to 3 different positions to the left of where the original values are located.

Press 2nd + (STAT) F2 [EDIT] to access the data lists. To copy the contents of one list to another list; for example, to copy the contents of L1 to L3, use ▲ and ► to move the cursor so that L3 is highlighted. Press ALPHA 7 (L) 1 ENTER.	L1 L2 L3 3 1999 2541 -------- 2000 3081 2001 3615 2002 4157 2003 4703 2004 5242 L3 =L1 < > NAMES " OPS ►

NOTE: This first step shown above is not necessary, but it will save you the time it takes to re-enter the input data if you make a mistake. Also, it is not necessary to first clear L3. However, if you want to do so, have the symbols L3 highlighted and press CLEAR ENTER.

To align the input data as the number of years past 1999, first press the arrow keys (◄ and ▲) so that L1 is highlighted. Tell the TI-86 to subtract 1999 from each number in L1 with ALPHA 7 (L) 1 – 1999.	L1 L2 L3 1 1999 2541 1999 2000 3081 2000 2001 3615 2001 2002 4157 2002 2003 4703 2003 2004 5242 2004 L1 =L1-1999 < > NAMES " OPS ►
Press ENTER. Instead of an actual year, the input now represents the number of years since 1999. Return to the home screen with 2nd EXIT (QUIT).	L1 L2 L3 1 0 2541 1999 1 3081 2000 2 3615 2001 3 4157 2002 4 4703 2003 5 5242 2004 L1(1) =0 < > NAMES " OPS ►
Find the linear function by pressing 2nd ENTER (ENTRY) as many times as needed until you see the linear fit instruction. To enter this function in a different location, say y2, press ◄ and 2.	LinR L1,L2,y2
Press ENTER and then press GRAPH F1 [y(x)=] to see the function pasted in the y2 location. *Note:* If you want the aligned function to be in y1, do not replace y1 with y2 before pressing ENTER to find the equation.	Plot1 Plot2 Plot3 \y1=-1077663.5809525… \y2=2538.9047619047+… y(x)= WIND ZOOM TRACE GRAPH x y INSf DELf SELCT►
To graph this equation on a scatter plot of the aligned data, first turn off the function in y1 (see page B-15 of this *Guide*). Press EXIT F3 [ZOOM] MORE F5 [ZDATA].	

If you now want to find the linear function that best fits the modified tax data using the input data aligned another way, say as the number of years after 1900, first return to the data lists with 2nd + (STAT) F2 [EDIT] and highlight L1.

Add 99 to each number currently in L1 with ALPHA 7 (L) 1 + 99 ENTER. Instead of an actual year, the input now represents the number of years since 1900.	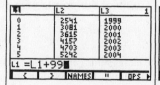 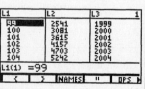

- There are many ways that you can enter the aligned input into L1. One method that you may prefer is to start over from the beginning. Replace L1 with the contents of L3 by highlighting L1 and pressing ⟦ALPHA⟧ 7 (L) 3 ⟦ENTER⟧. Once again highlight the name L1 and subtract 1900 from each number in L1 with ⟦ALPHA⟧ 7 (L) 1 ⟦-⟧ 1900.

On the home screen, find the linear function for the aligned data by pressing ⟦2nd⟧ ⟦ENTER⟧ (ENTRY) until you see the linear regression instruction. To enter this new equation in a different location, say y3, press ⟦◄⟧ 3 ⟦ENTER⟧. Press ⟦GRAPH⟧ ⟦F1⟧ [y(x)=] to see the function pasted in the y3 location.

```
Plot1 Plot2 Plot3
\y1=-1077663.5809525…
\y2=2538.9047619047+…
\y3=-50957.866666669…

y(x)= WIND ZOOM TRACE GRAPH
  x    y   INSf DELf SELCT
```

- To graph this equation on a scatter plot of the aligned data, first turn off the other functions and then press ⟦EXIT⟧ ⟦F3⟧ [ZOOM] ⟦MORE⟧ ⟦F5⟧ [ZDATA].

- Remember, if you have aligned the data, the input value at which you evaluate the function may not be the value given in the question you are asked. You can use any of the equations to evaluate function values.

1.4.3b USING A MODEL FOR PREDICTIONS   You can use one of the methods described in Section 1.1.1e or Section 1.1.1f of this *Guide* to evaluate the linear function at the indicated input value. Remember, if you have aligned the data, the input value at which you evaluate the function may not be the value given in the question you are asked.

CAUTION: Remember that you should always use the full model, *i.e.*, the function you pasted in the y(x)= list, not a rounded equation, for all computations.

Using the function in y1 (the input is the year), in y2 (the input is the number of years after 1999), or in y3 (the input is the number of years after 1900), we predict that the tax owed in 2006 is approximately $6322.

```
y1(2006)
            6321.5047618
y2(7)
            6321.5047619
y3(106)
           6321.50476191
```

You can also predict the tax in 2006 using the TI-86 table (with ASK chosen in TBLSET) and any of the 3 models found in the previous section of this *Guide*. As seen to the right, the predicted tax is approximately $6322.

```
    x        y1
  2006    6321.5047…

y1=6321.5047618
TBLST SELCT   x    y
```

1.4.3c NAMING AND STORING DATA   You can name data (either input, output, or both) and store it in the calculator memory for later recall. You may or may not want to use this feature. It will be helpful if you plan to use a large data set several times and do not want to reenter the data each time.

To illustrate the procedure, let's name and store the modified tax output data that was entered in Section 1.4.2b.

Press ⟦2nd⟧ ⟦EXIT⟧ (QUIT) to return to the home screen. You can view any list from the STAT EDIT mode (where the data is originally entered) or from the home screen. View the modified tax data in L2 by pressing ⟦ALPHA⟧ 7 (L) 2 ⟦ENTER⟧.

```
L2
{2541 3081 3615 4157…
```

> Pressing ▶ allows you to scroll through the list to see the portion that is not displayed. To store this data with the name TAX, press ALPHA 7 (L) 2 STO▶ T A X ENTER.

CAUTION: Do not store data to a name that is routinely used by the TI-86. Such names are ANS, MATH, LOG, MODE, A, B, L1, L2, ..., L6, and so forth. Note that if you use a single letter as a name, this might cause one or more of the programs to not execute properly.

1.4.3d RECALLING STORED DATA The data you have stored remains in the memory of the TI-86 until you delete it using the instructions given in Section 1.4.3e of this *Guide*. When you wish to use the stored data, recall it to one of the lists L1, L2, ..., L5. We illustrate with the list named TAX, which we store in L2. Press 2nd − (LIST) and under NAMES, find TAX. (You may need to press MORE until you see the list you want to recall.)

> Press the F-key corresponding to the location of the list. Press STO▶ 7 (L) ALPHA 2 ENTER. List L2 now contains the TAX data.

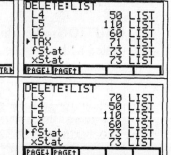

1.4.3e DELETING USER-STORED DATA You do not need to delete any data lists unless your memory is getting low or you just want to. For illustration purposes, we delete the TAX list.

> Press 2nd 3 (MEM) F2 [DELET]. Next press F4 [LIST] and use ▼ to move the cursor opposite TAX.
>
> Press ENTER. To delete another list, use ▼ or ▲ to move the cursor opposite that list name and press ENTER. Exit this screen with 2nd EXIT (QUIT) when finished.

WARNING: Be careful when in the DELETE menu. Once you delete something, it is gone from the TI-86 memory and cannot be recovered. If you mistakenly delete one of the lists L1, L2, ..., L5, you need to redo the statistical setup as indicated on page B-1 of this *Guide*.

1.4.4 WHAT IS "BEST FIT"? It is important to understand the method of least squares and the conditions necessary for its use if you intend to find the equation that best-fits a set of data. You can explore the process of finding the line of best fit with program LSLINE. (Program LSLINE is given in the *TI-86 Program Appendix*.) For your investigations of the least-squares process with this program, it is better to use data that is not perfectly linear and data for which you do *not* know the best-fitting line.

  We use the data in the table below to illustrate program LSLINE, but you may find it more interesting to input some other data[4].

---

[4] This program works well with approximately 5 data points. Interesting data to use in this illustration are the height and weight, the arm span length and the distance from the floor to the navel, or the age of the oldest child and the number of years the children's parents have been married for 5 randomly selected persons.

x (input in L1)	1	3	7	9	12
y (output in L2)	1	6	10	16	20

Before using program LSLINE, clear all functions from the y(x)= list, turn on Plot1 by darkening the name Plot1 on the y(x)= screen, and enter your data in lists L1 and L2. (If Plot 1 is not turned on, the program will not execute properly.) Next, draw a scatter plot with GRAPH F3 [ZOOM] MORE F5 [ZDATA].

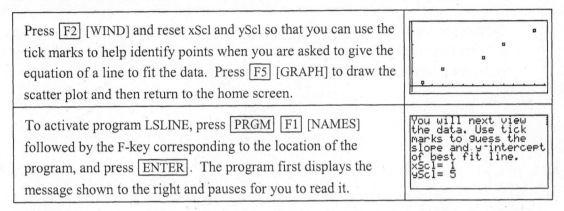

Press F2 [WIND] and reset xScl and yScl so that you can use the tick marks to help identify points when you are asked to give the equation of a line to fit the data. Press F5 [GRAPH] to draw the scatter plot and then return to the home screen.

To activate program LSLINE, press PRGM F1 [NAMES] followed by the F-key corresponding to the location of the program, and press ENTER. The program first displays the message shown to the right and pauses for you to read it.

NOTE: While the program is calculating, there is a small vertical line in the upper-right hand corner of the graphics screen that is dashed and "wiggly". This program pauses several times during execution for you to view the screen. Whenever the program pauses, the small line is "still" and you should press ENTER to resume execution after you have looked at the screen.

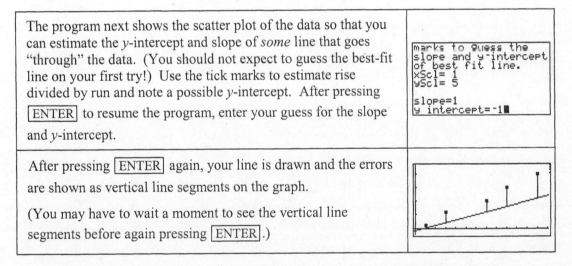

The program next shows the scatter plot of the data so that you can estimate the y-intercept and slope of *some* line that goes "through" the data. (You should not expect to guess the best-fit line on your first try!) Use the tick marks to estimate rise divided by run and note a possible y-intercept. After pressing ENTER to resume the program, enter your guess for the slope and y-intercept.

After pressing ENTER again, your line is drawn and the errors are shown as vertical line segments on the graph.

(You may have to wait a moment to see the vertical line segments before again pressing ENTER.)

Next the sum of squared errors, SSE, is displayed for your line. Decide whether you want to move the y-intercept of the line or change its slope to improve the fit to the data.

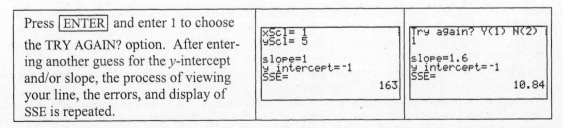

Press ENTER and enter 1 to choose the TRY AGAIN? option. After entering another guess for the y-intercept and/or slope, the process of viewing your line, the errors, and display of SSE is repeated.

If the new value of SSE is smaller than the SSE for your first guess, you have improved the fit.

When you feel an SSE value close to the minimum value is found, enter 2 at the TRY AGAIN? prompt. The program then overdraws the line of best fit on the graph and shows the errors for the line of best fit.

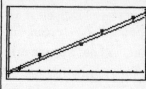
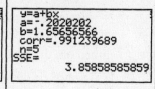

```
y=a+bx
a=-.2020202
b=1.65656566
corr=.991239689
n=5
SSE=
          3.85858585859
```

The program ends by displaying the coefficients a and b of the best-fit line $y = ax + b$ as well as the minimum SSE. Press ENTER to end the program. Use program LSLINE to explore[5] the method of least squares[6] that the TI-86 uses to find the line of best fit.

---

[5] Program LSLINE is for illustration purposes only. Actually finding the line of best fit for a set of data should be done according to the instructions in Section 1.5.7 of this *Guide*.

[6] Two articles that further explain "best-fit" are H. Skala, "Will the Real Best Fit Curve Please Stand Up?" Classroom Computer Capsule, *The College Mathematics Journal*, vol. 27, no. 3, May 1996 and Bradley Efron, "Computer-Intensive Methods in Statistical Regression," *SIAM Review*, vol. 30, no. 3, September 1988.

Chapter 2   Ingredients of Change: Nonlinear Models

 ## 2.1 Exponential Functions and Models

As we begin to consider functions that are not linear, it is very important that you be able to draw scatter plots, find numerical changes in output data, and recognize the underlying shape of the basic functions to be able to identify which function best models a particular set of data. Finding the model is only a means to an end – being able to use mathematics to describe the changes that occur in real-world situations.

2.1.1a ENTERING EVENLY SPACED INPUT VALUES   When an input list consists of many values that are the same distance apart, there is a calculator command that will generate the list so that you do not have to type the values in one by one. The syntax for this sequence command is *seq(formula, variable, first value, last value, increment)*. When entering years that differ by 1, the formula is the same as the variable and the increment is 1. Any letter can be used for the variable -- we choose to use x. We illustrate the sequence command use to enter the input for the population data in Table 2.2 in Section 2.1 of *Calculus Concepts*.

Year	1994	1995	1996	1997	1998	1999	2000	2001	2002	2003
Population	7290	6707	6170	5677	5223	4805	4420	4067	3741	3442

Clear any old data from lists L1 and L2. To enter the input data, position the cursor in the first list so that L1 is darkened. Generate the list of years beginning with 1994, ending with 2003, and differing by 1 with [2nd] [−] (LIST) [F5] [OPS] [MORE] [F3] [seq(] [x-VAR] [,] [x-VAR] [,] 1994 [,] 2003 [,] 1 [)].

Press [ENTER] and the sequence of years is pasted into L1. Manually enter the output values in L2.

2.1.1b FINDING PERCENTAGE DIFFERENCES   When the input values are evenly spaced, use program DIFF to compute the percentage differences in the output data. If the data are perfectly exponential (that is, every data point falls on the exponential function), the percentage change (which in this case would equal the percentage differences) in the output is constant. If the percentage differences are close to constant, this is an indication that an exponential model *may* be appropriate.

Return to the home screen. Find the percentage differences in the population data with program DIFF. Press [PRGM] [F1] [NAMES] and then press the F-key that is under program DIFF.

Press [ENTER], choose [F1] [Yes] to continue, and at the Choice? prompt, press [F3] [%] to generate the list of percentage differences. Press [F4] [Quit] to exit the program.

The list will not scroll to the right because it was generated from within a program. However, the program has stored the percentage differences in list L5.

B-31

Press 2nd + (STAT) F2 [EDIT] and use ▶ to view the percentage differences in list L5. Scroll down L5 with ▼ to see all 9 values. The percentage differences are close to constant, so an exponential function will probably fit these data.

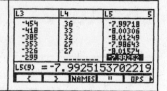

2.1.2 FINDING AN EXPONENTIAL FUNCTION TO MODEL DATA Use your calculator to find an exponential equation to model the small town population data. The exponential function we use is of the form $y = ab^x$. It is very important that you align large numbers (such as years) whenever you find an exponential model. If you don't align to smaller values, the TI-86 may return an error message or an incorrect function due to computation errors. We continue to use the data in Table 2.2 of the text.

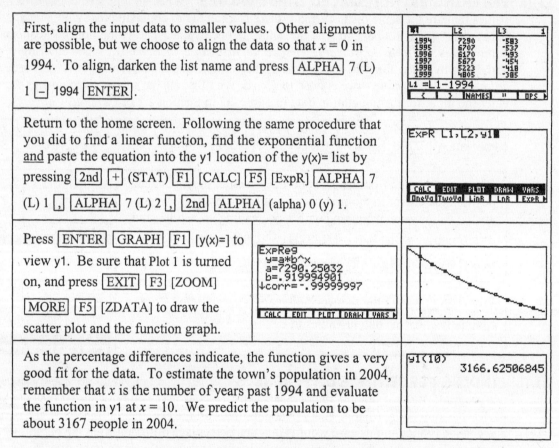

First, align the input data to smaller values. Other alignments are possible, but we choose to align the data so that $x = 0$ in 1994. To align, darken the list name and press ALPHA 7 (L) 1 − 1994 ENTER.

Return to the home screen. Following the same procedure that you did to find a linear function, find the exponential function and paste the equation into the y1 location of the y(x)= list by pressing 2nd + (STAT) F1 [CALC] F5 [ExpR] ALPHA 7 (L) 1 , ALPHA 7 (L) 2 , 2nd ALPHA (alpha) 0 (y) 1.

Press ENTER GRAPH F1 [y(x)=] to view y1. Be sure that Plot 1 is turned on, and press EXIT F3 [ZOOM] MORE F5 [ZDATA] to draw the scatter plot and the function graph.

As the percentage differences indicate, the function gives a very good fit for the data. To estimate the town's population in 2004, remember that x is the number of years past 1994 and evaluate the function in y1 at $x = 10$. We predict the population to be about 3167 people in 2004.

- Do not confuse the percentage differences found from the data with the percentage change for the exponential function. The constant percentage change in the exponential function $y = ab^x$ is $(b - 1)100\%$. The constant percentage change for the function is a single value whereas the percentage differences calculated from the data are many different numbers.

2.2 Logarithmic Functions and Models

In this section we consider another function that can be used to fit data – the log function. This function is the inverse function for the exponential function discussed in the previous section. We recognize when to use this function by considering the behavior of the data rather than a numerical test involving differences.

2.2.1 FINDING A LOG FUNCTION TO MODEL DATA   Use your calculator to find a log equation of the form $y = a + b \ln x$. We illustrate finding this function with the air pressure and altitude data in Table 2.5 of Section 2.2 of *Calculus Concepts*.

Air pressure (inches of mercury)	13.76	5.56	2.14	0.82	0.33
Altitude (thousands of feet)	20	40	60	80	100

Clear the data that is currently in lists L1 and L2. Enter the air pressure data in L1 and the altitude data in L2. (If you wish, clear list L3, but it is not necessary to do so.) Remember to clear the y(x)= list and turn Plot 1 on.

A scatter plot of the data drawn with GRAPH F3 [ZOOM] MORE F5 [ZDATA] shows the slow decline that often indicates a log model.

As when modeling linear and exponential functions, find and paste the log equation into the y1 location of the y(x)= list by pressing 2nd + (STAT) F1 [CALC] F4 [LnR] ALPHA 7 (L) 1 ⌐ , ALPHA 7 (L) 2 ⌐ , 2nd ALPHA (alpha) 0 (y) 1.

Press y(x)= to view the function. Either press GRAPH F3 [ZOOM] MORE F5 [ZDATA] or F5 [GRAPH] to view the func-tion on the plot of the data. It appears to be a great fit.

2.2.2 AN INVERSE RELATIONSHIP   Exponential and log functions are inverse functions. If you find the inverse function for the log equation that models altitude as a function of air pressure (given in the section above), it will be an exponential function. We briefly explore this concept using the TI-86. Leave the log function found in the previous section ($y = 76.174 - 21.331 \ln x$) in the y1 location of the y(x)= list because we use it again in this investigation.

Find the exponential equation that models air pressure as a function of altitude and paste it into the y2 location by pressing 2nd + (STAT) F1 [CALC] F5 [ExpR] ALPHA 7 (L) 2 ⌐ , ALPHA 7 (L) 1 ⌐ , 2nd ALPHA (alpha) 0 (y) 2.

Note that L2 is typed first!

Press ENTER. Then press GRAPH F1 [y(x)=] and turn Plot 1 off.

Enter the composite function y1 ∘ y2 (or y2 ∘ y1) in the y3 location. We use the Composition Property of Inverse Functions.

What is returned for y3 in the table is close to the value of x that was input.

Press ▶ to see the y3 column.

- There is a slight difference in the exponential function and the true inverse function for y1 because of rounding and because the exponential function in y2 was not found algebraically but from fitting an equation to data points.

You can also graphically verify that y1 and y2 are inverse functions. Clear or turn off the composite function in y3 and turn off y2. Return to the home screen of the TI-86.

Have the TI-86 draw the graph of the inverse of y1 with GRAPH MORE F2 [Draw] MORE MORE MORE F3 [DrawInv] 2nd ALPHA (alpha) 0 (y) 1 ENTER . Press GRAPH F1 [y(x)=], turn y2 on, and press EXIT F5 [GRAPH].

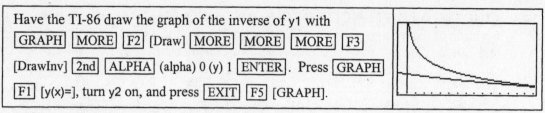

- You see the same graph, indicating that y2 and the inverse function either are the same function or close enough to not be visually distinguished.

2.3 Logistic Functions and Models

This section introduces the logistic function that can be used to describe growth that begins as exponential and then slows down to approach a limiting value. Logistic functions are often in applications to the spread of a disease or virus. The TI-86's random number generator can be used to collect data to illustrate such a situation.

2.3.1 GENERATING RANDOM NUMBERS

Imagine all the real numbers between 0 and 1, including 0 but not 1, written on identical slips of paper and placed in a hat. Close your eyes and draw one slip of paper from the hat. You have just chosen a number "at random."

Your calculator doesn't offer you a random choice of all real numbers between 0 and 1, but it allows you to choose, *with an equal chance of obtaining each one*, any of 10^{12} different numbers between 0 and 1 with its random number generator called rand.

Before using your calculator's random number generator for the first time, you should "seed" the random number generator. (This is like mixing up all the slips of paper in the hat.)

Pick some *whole* number, <u>not</u> the one shown on the right, and store it as a "seed" with *your number* STO▸ 2nd X (MATH) F2 [PROB] F4 [rand] ENTER . Everyone needs to have a different seed, or the number choices will not be random.	
Enter rand again, and press ENTER several times.

Your list of random numbers should be different from the one on the right. Notice that all numbers are between 0 and 1. | |

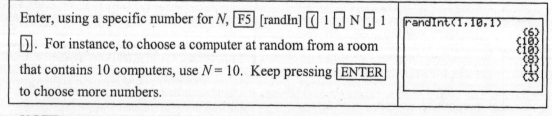

We usually count with the numbers 1, 2, 3, ... and so forth, not with decimal values. So, the TI-86 has a key called randInt that chooses a whole number between, and including, 1 and N.

Enter, using a specific number for N, F5 [randIn] (1 , N , 1) . For instance, to choose a computer at random from a room that contains 10 computers, use N = 10. Keep pressing ENTER to choose more numbers.	

NOTE: If you want to generate a list of five random numbers, change the last 1 in the above instruction to 5; if you want 10 random numbers, change it to 10, and so forth.

As, previously stated, your list of numbers should be different from those shown above. All values will be between 1 and 10. Note that it is possible to obtain the same value more than once. This corresponds to each slip of paper being put back in the hat after being chosen and the numbers mixed well before the next number is drawn.

2.3.2a FINDING A LOGISTIC FUNCTION TO MODEL DATA   You can use your calculator to find a logistic function of the form $y = \dfrac{L}{1 + Ae^{-Bx}} + d$. Unlike the other functions we have fit to data, the logistic equations that you obtain will be slightly different from those given in the text answer key and different from logistic equations found with another model calculator because the TI-86 equation has a built-in vertical shift (d). (The logistic equations in *Calculus Concepts* were found using a TI-83 that does not have a built-in vertical shift.) As you did when finding an exponential equation for data, large input values must be aligned or an error or possibly an incorrect answer could be the result.

Note that the TI-86 finds a "best-fit" logistic function rather than a logistic function with a limiting value L such that no data value is ever greater than L. We illustrate finding a logistic function using the data in Table 2.8 of Section 2.3 in *Calculus Concepts*.

Time (hours after 8 A.M.)	0	1	2	3	4	5
Total number of infected computers	1	2	4	6	8	9

Clear any old data.   Delete any functions in the y(x)= list and turn on Plot 1.   Enter the data in the above table with the hours after 8 A.M. in list L1 and the number of computers in list L2:

Construct a scatter plot of the data. An inflection point is indicated, and the data appear to approach limiting values on the left and on the right. A logistic model seems appropriate.	
To find the logistic function and paste the equation into the y1 location of the y(x)= list, press [2nd] [+] (STAT) [F1] [CALC] [MORE] [F3] [LgstR] [ALPHA] 7 (L) 1 [,] [ALPHA] 7 (L) 2 [,] [2nd] [ALPHA] (alpha) 0 (y) 1.	`LgstR L1,L2,y1`

Press [ENTER]. This equation takes longer to generate than the other equations. Press [GRAPH] [F5] [GRAPH] to view the function on the scatter plot.	`LogisticReg`
	`y=a/(1+be^(cx))+d`
	`n=6`
	`tolMet=1`
	`PRegC=`
	`{9.90915115219 9.635...`

NOTE: The TI-86 uses the variable a for the value that the text calls L, b for the value the text calls A, and c for the value the text calls B. The variable d represents the vertical shift. You can see the values of these parameters by pressing [▶] to view the list {a b c d} that is shown in the middle box above or you can press [▶] and see them in the equation in the y(x)= list.

The limiting value, in this context, is about 10 computers. The two horizontal asymptotes are $y = 0$ (the line lying along the *x*-axis) and the line $y \approx 10$. To see the upper asymptote drawn on a graph of the data, enter 10 in y2 and press [GRAPH].	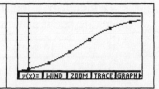

- Provided the input values are evenly spaced, program DIFF might be helpful when you are trying to determine if a logistic model is appropriate for certain data. If the first differences (in list L3 after running program DIFF) begin small, peak in the middle, and end small, this is an indication that a logistic model may provide a good fit to the data.

2.3.2b RECALLING MODEL PARAMETERS   Rounding function parameters can often lead to incorrect or misleading results. You may find that you need to use the complete values of the coefficients after you have found a function that best fits a set of data. It would be tedious to copy all these digits in a long decimal number into another location of your calculator. You don't have to because you can recall any parameter found after you use one of the regressions.

We illustrate these ideas by recalling the parameters for the logistic function found in the previous section of this *Guide*. These parameters can be recalled as a list or as individually.

To recall the entire list, which contains the coefficients listed in alphabetical order, press ☐2nd☐ ☐+☐ (STAT) ☐F5☐ [VARS] ☐MORE☐ ☐MORE☐ ☐MORE☐ ☐MORE☐ ☐F2☐ [PRegC] ☐ENTER☐. Recall that ▶ scrolls the list to the right for viewing all the values.	PRegC …1.26538130244 .8870… CALC EDIT PLOT DRAW **VARS** Med PRegC QrtI1 QrtI3 toIMe

Using the TI-86 Catalog: You can obtain most of the calculator commands by pressing ☐2nd☐ ☐CUSTOM☐ [CATLG-VARS], choosing the appropriate F-key (use ☐F1☐ [CATLG] or ☐F2☐ [ALL] if you are not sure what type of item you are trying to find), pressing the key corresponding to the first letter of the name of the item, using ☐▼☐ until the cursor is opposite the item, and then pressing ☐ENTER☐ to copy the item to wherever you had the cursor before entering the catalog.   For example, to type PRegC, press ☐2nd☐ ☐CUSTOM☐ [CATLG-VARS] ☐MORE☐ ☐MORE☐ ☐F4☐ [STAT] ☐▼☐ ☐ENTER☐.

To recall the value of a in $y = a/(1+ be^{\wedge}(-cx)) + d$, first follow the directions given above to type PRegC on the home screen. Then, because a is first in the list, press ☐(☐ 1 ☐)☐ ☐ENTER☐. To recall b, change 1 to 2 because b is second in the list, and so forth.	PRegC(1)
	          785.473942636
	PRegC(2)
	          1504.56498821
	PRegC(3)
	          -1.26538130244
	PRegC(4)■

- This same procedure applies to any equation you have found using the STAT CALC menu. Observe the name of the list that appears when you find the equation and call up that list to find the coefficients in the function. If no list appears (such as when finding a linear or exponential function), the coefficients are given as a and b in the CATLG STAT menu.

WARNING: Once a different function is found, no parameters from a previous function that you have fit to data remain stored in the calculator's memory.

2.3.3 VERTICALLY SHIFTING DATA   Because the TI-86 has a built-in vertical shift for the logistic function, this section of the *Guide* is not necessary for logistic models. Thus, instead of the investment club data in Table 2.13 in Example 1 of Section 2.3 of *Calculus Concepts* (which uses a logistic function), we use the exponential data in Table 2.11 on page **xxx** of the text to see if a vertical shift improves the fit.

x	0	2	4	6	8
f(x)	101	104	116	164	356

First, clear your lists, and then enter the data in the above table in L1 and L2, respectively.

Draw a scatter plot of the data and notice that the shape indicates an exponential model may be appropriate.	L1 L2 L3 2 0 101 -------- 2 104 4 116 6 164 8 356 L2(6) = { } NAMES " OPS ▶

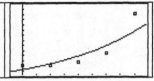

However, when we find an exponential function, it is a poor fit for the data. Also notice how the scatter plot appears to level off as the input approaches zero.

This situation calls for shifting the data vertically so that it approaches a lower asymptote of $y = 0$, which is what the exponential function in the TI-86 has as its lower asymptote for an increasing exponential curve. There are many choices of how to shift the data. We choose to subtract 100 from each output data value.

Highlight L2 at the top of the second column, type L2 − 100, and then press ENTER . Now fit an exponential function to the shifted data.	
The shifted function is therefore $g(x) = 2x + 100$. Add 100 to the function in y1.	
If you wish to graph g on the original data, add 100 to each output value by highlighting L2 at the top of the second column, typing L2 − 100, and then pressing ENTER .	
Draw the graphs of g and the original data with GRAPH F3 [ZOOM] MORE F5 [ZDATA]. The shifted function provides an excellent fit to the data.	

NOTE: If you fit more than one shifted exponential function and are unsure which of them better fits the data, zoom in on each near the "ends" of each graph to see which function seems to best follow the pattern of the data.

2.4 Polynomial Functions and Models

You will in this section learn how to fit functions that have the familiar shape of a parabola or a cubic to data. Using your calculator to find these equations involves the same procedure as when using it to fit linear, exponential, log, or logistic functions.

2.4.1a FINDING SECOND DIFFERENCES When the input values are evenly spaced, you can use program DIFF to quickly compute second differences in the output values. If the data are perfectly quadratic (that is, every data point falls on a quadratic function), the second differences in the output values are constant. When the second differences are close to constant, a quadratic function *may* be appropriate for the data.

We illustrate these ideas with the roofing jobs data given on the first page of Section 2.4 in *Calculus Concepts*. We align the input data so that 1 = January, 2 = February, etc. Clear any old data and enter these data in lists L1 and L2:

Month of the year	1	2	3	4	5	6
Number of roofing jobs	117	140	224	368	575	842

The input values are evenly spaced, so we can see what information is given by viewing the second differences.

Run program DIFF to see the first differences (also in list L3), the second differences (also in list L4), and the percentage differences (also in list L5). The second differences are close to constant, so a quadratic function may give a good fit for these data.	Choice? 1st differences in L3 {23 84 144 207 267} 2nd differences in L4 {61 60 63 60} Percent diffs in L5 {19.6581196581 60 64... | 1st | 2nd | % | Quit |
Construct a scatter plot of the data. (Don't forget to clear the y(x)= list of previously-used equations and to turn on Plot 1.) The shape of the data confirms the numerical investigation result that a quadratic function is appropriate.	

2.4.1b FINDING A QUADRATIC FUNCTION TO MODEL DATA

Use your TI-86 to find a quadratic function to model the roofing jobs data. The TI-86's quadratic function is of the form $y = ax^2 + bx + c$ and is fit with the command P2Reg.

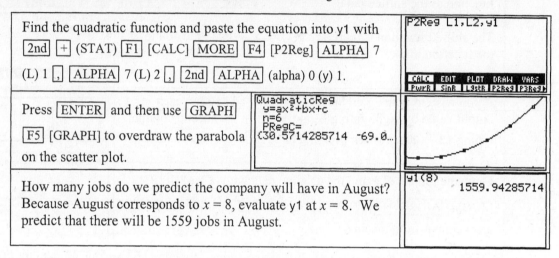

Find the quadratic function and paste the equation into y1 with [2nd] [+] (STAT) [F1] [CALC] [MORE] [F4] [P2Reg] [ALPHA] 7 (L) 1 [,] [ALPHA] 7 (L) 2 [,] [2nd] [ALPHA] (alpha) 0 (y) 1.	P2Reg L1,L2,y1 | CALC | EDIT | PLOT | DRAW | VARS | | PwrR | SinR | LgstR | P2Reg | P3Reg |
Press [ENTER] and then use [GRAPH] [F5] [GRAPH] to overdraw the parabola on the scatter plot.	QuadraticReg y=ax²+bx+c n=6 PRegC= {30.5714285714 -69.0...
How many jobs do we predict the company will have in August? Because August corresponds to $x = 8$, evaluate y1 at $x = 8$. We predict that there will be 1559 jobs in August.	y1(8) 1559.94285714

2.4.2 FINDING A CUBIC FUNCTION TO MODEL DATA

Whenever a scatter plot of data shows a single change in concavity, we are limited to fitting either a cubic or logistic function. If one or two limiting values are apparent, use the logistic equation. Otherwise, a cubic function should be considered. The TI-86's cubic function is of the form $y = ax^3 + bx^2 + cx + d$ and is accessed using the P3Reg key.

We illustrate finding a cubic function with the data in Table 2.18 in Example 3 of Section 2.4 in *Calculus Concepts*. The data give the average price in dollars per 1000 cubic feet of natural gas for residential use in the U.S. for selected years between 1980 and 2000.

Year	1980	1982	1985	1990	1995	1998	2000
Price (dollars)	3.68	5.17	6.12	5.80	6.06	6.82	7.71

First, clear your lists, and then enter the data. Next, align the input data so that x represents the number of years since 1980. (We do not have to align here, but we do so in order to have smaller coefficients in the cubic function.)	L1 | L2 | L3 | 1 0 | 3.68 | ---- 2 | 5.17 5 | 6.12 10 | 5.80 15 | 6.06 18 | 6.82 L1 =L1-1980 | { | } | NAMES | " | OPS |
Clear any functions in the y(x)= list, and draw a scatter plot of these data. Notice that a concavity change is evident, but there do not appear to be any limiting values. Thus, a cubic model is appropriate to fit the data.	

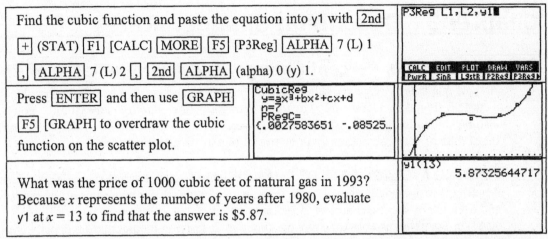

Find the cubic function and paste the equation into y1 with 2nd + (STAT) F1 [CALC] MORE F5 [P3Reg] ALPHA 7 (L) 1 , ALPHA 7 (L) 2 , 2nd ALPHA (alpha) 0 (y) 1.		P3Reg L1,L2,y1
Press ENTER and then use GRAPH F5 [GRAPH] to overdraw the cubic function on the scatter plot.	CubicReg y=ax³+bx²+cx+d n=7 PRegC= {.0027583651 -.08525...	
What was the price of 1000 cubic feet of natural gas in 1993? Because *x* represents the number of years after 1980, evaluate y1 at *x* = 13 to find that the answer is \$5.87.		y1(13) 5.87325644717

Part *b* of Example 3 asks you to find when, according to the model, the price of 1000 cubic feet of natural gas first exceed \$6. You can find the answer by using the SOLVER to find the solution to y1 = 6. However, since we already have a graph of this function, the following may be easier. Note that if you do use the SOLVER, you should first TRACE the graph to estimate a guess for *x* that you need to enter in the SOLVER.

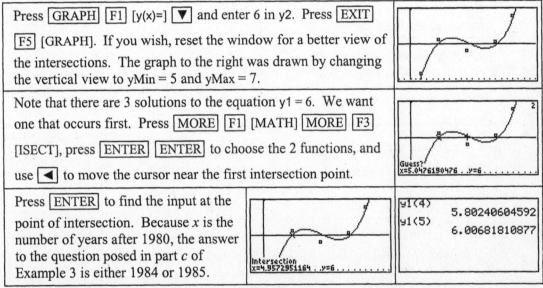

Press GRAPH F1 [y(x)=] ▼ and enter 6 in y2. Press EXIT F5 [GRAPH]. If you wish, reset the window for a better view of the intersections. The graph to the right was drawn by changing the vertical view to yMin = 5 and yMax = 7.	
Note that there are 3 solutions to the equation y1 = 6. We want one that occurs first. Press MORE F1 [MATH] MORE F3 [ISECT], press ENTER ENTER to choose the 2 functions, and use ◄ to move the cursor near the first intersection point.	Guess? x=5.0476190476 . .y=6 . . . .
Press ENTER to find the input at the point of intersection. Because *x* is the number of years after 1980, the answer to the question posed in part *c* of Example 3 is either 1984 or 1985.	Intersection x=4.9572951164 . .y=6 . . . . y1(4) 5.80240604592 y1(5) 6.00681810877

The price had not *exceeded* \$6 in 1984, so the answer to the question is 1985. Read carefully!

Notice that the natural gas prices could have also been modeled with a cubic equation by renumbering the input data so that *x* is the number of years after 1900. Return to the data and add 80 to each input value so that L1 represents the number of years after 1900 rather than the number of years after 1980. Now find a cubic function to fit these aligned data.

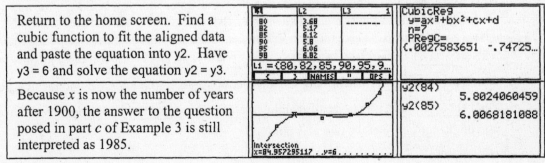

Return to the home screen. Find a cubic function to fit the aligned data and paste the equation into y2. Have y3 = 6 and solve the equation y2 = y3.	L1 L2 L3 1 80 3.68 ------- 82 5.17 85 6.12 90 5.8 95 6.06 98 6.82 L1 ={80,82,85,90,95,9... CubicReg y=ax³+bx²+cx+d n=7 PRegC= {.0027583651 -.74725...
Because *x* is now the number of years after 1900, the answer to the question posed in part *c* of Example 3 is still interpreted as 1985.	Intersection x=84.957295117 . .y=6 . . . . y2(84) 5.8024060459 y2(85) 6.0068181088

Chapter 3   Describing Change: Rates

3.1 Change, Percent Change, and Average Rates of Change

As you calculate average and other rates of change, remember that every numerical answer in a context should be accompanied by units telling how the quantity is measured. You should also be able to interpret each numerical answer. It is only through their interpretations that the results of your calculations will be useful in real-world situations.

3.1.1a FINDING AVERAGE RATE OF CHANGE   Finding an average rate of change using a function is just a matter of evaluating the equation at two different values of the input variable and dividing by the difference of those input values.

We illustrate this concept using the function describing the population density of Nevada from 1960 through 2000 that is given in Example 3 of Section 3.1 of *Calculus Concepts*.

The population density referred to above is given by the function $p(t) = 0.1536(1.04892^t)$ people per square mile where t is the number of years after 1900. Press GRAPH F1 [y(x)=], clear any functions, turn Plot 1 off, and enter this function in y1 using x as the input variable.

To find the average rate of change of the population density from 1960 through 1980, first realize that 1960 corresponds to $x = 60$ and 1980 corresponds to $x = 80$. Then, recall that the average rate of change of p between 1960 and 1980 is given by the quotient $\frac{p(80)-p(60)}{80-60}$.

Return to the home screen and type the quotient $\frac{y1(80)-y1(60)}{80-60}$. Remember to enclose both the numerator and denominator of the fraction in parentheses. Finding this quotient in a single step avoids having to round intermediate calculation results.	(y1(80)-y1(60))/(80-6 0) .215684221348

Recall that rate of change units are output units per input units. On average, Nevada's population density increased about 0.22 person per square mile per year between 1960 and 1980.

To find the average rate of change between 1980 and 2000, recall the last expression with 2nd ENTER (ENTRY) and replace 60 with 100 in two places. Press ENTER.	(y1(80)-y1(60))/(80-6 0) .215684221348 (y1(80)-y1(100))/(80- 100) .560616273917

NOTE: If you have many average rates of change to calculate, you could put the average rate of change formula in the graphing list: y2 = (y1(B) – y1(A))/(B – A). Of course, you need to have a function in y1. Then, on the home screen, store the inputs of the two points in A and B with 80 STO▶ A 2nd . (:) ALPHA 100 STO▶ B ENTER. All you need do next is type y2 and press ENTER. Then, store the next set of inputs into A and/or B and recall y2, using 2nd ENTER (ENTRY) to recall each instruction. Press ENTER and you have the average rate of change between the two new points. Try it!

3.1.1b CALCULATING PERCENTAGE CHANGE   You can find percentage changes using data either by the formula or by using program DIFF. To find a percentage change from a function instead of data, you should use the percentage change formula. We again use the population density of Nevada function in Example 3 of Section 3.1 of *Calculus Concepts* to illustrate.

Have $p(x) = 0.1536(1.04892^t)$ in y1. The percentage change in the population density between 1960 and 2000 is given by the formula $\dfrac{p(100) - p(60)}{p(60)} \cdot 100\% = \dfrac{y1(100) - y1(60)}{y1(60)} \cdot 100\%$. Recall that if you type in the entire quotient, you avoid rounding errors.	`(y1(100)-y1(60))/(y1(`
	`60)`
	`           5.75607794432`
	`Ans*100`
	`           575.607794432`

The population density of Nevada increased by about 575.6% between 1960 and 2000.

3.2 Instantaneous Rates of Change

We first examine the principle of local linearity, which says that if you are close enough, the tangent line and the curve are indistinguishable. We also explore two methods for using the calculator to draw a tangent line at a point on a curve.

3.2.1a MAGNIFYING A PORTION OF A GRAPH   The ZOOM menu of your calculator allows you to magnify any portion of a graph. Consider the graph shown in Figure 3.7 in Section 3.2 of *Calculus Concepts*. The temperature model is $T(x) = -0.804x^2 + 11.644x + 38.114$ degrees Fahrenheit where x is the number of hours after 6 a.m.

Enter the temperature equation in y1 and draw the graph between 6 a.m. ($x = 0$) and 6 p.m. ($x = 12$) using GRAPH F3 [ZOOM] MORE F1 [ZFIT]. We now want to zoom and "box" in several points on the graph to see a magnified view.	
The first point we consider on the graph is point A with $x = 3$. Press F3 [ZOOM] F1 [Box] and use the arrow keys (◄, ▲, etc.) to move the cursor to the left of the curve close to where $x = 3$. (You may not have the same coordinates as those shown on the right.) Press ENTER to fix the lower left corner of the box.	`x=2.8571428571   y=67.353169629`
Use the arrow keys to move the cursor to the opposite corner of your "zoom" box. Point A should be close to the center of your box.	`x=4           y=70.073092385`
Press ENTER to magnify the portion of the graph that is inside the box. Look at the dimensions of the view you now see with EXIT [WIND]. Repeat the above process if necessary. The closer you zoom in on the graph, the more it looks like a line.	

Reset xMin to 0, xMax to 12, and redraw the graph of T with ZFIT. Repeat the zoom and box steps for point B with $x = 7.24$ and point C with $x = 9$. Note for each point how the graph of the parabola looks more like a line the more you magnify the view. These are illustrations of *local linearity*.

3.2.1b DRAWING A TANGENT LINE The TI-86 can draw a line tangent to the graph of a function at a specified point. To illustrate the process, we draw several tangent lines on the graph of T, the temperature function given in the previous section of this *Guide*.

Enter $T(x) = -0.804x^2 + 11.644x + 38.114$ in y1. Next, press F2 [WIND] and set xMin = 0, xMax = 12, yMin = 38, and yMax = 85. Press F5 [GRAPH]. On the graph screen, press MORE F1 [MATH] MORE MORE F1 [TANLN] 3 ENTER, and the tangent line is drawn at $x = 3$.

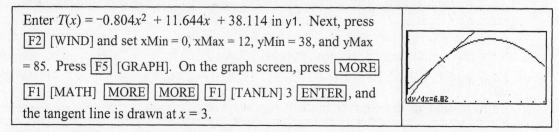

Note that the slope of the tangent line, symbolized by dy/dx, is displayed at the bottom of the screen. Compare this line with the zoomed-in view of the curve at point A (in the previous section). They actually are the same line if you are at point A!

We illustrate a slightly different method to draw the tangent line to the curve at point B where $x = 7.24$. Press GRAPH. Then press MORE F2 [DRAW] MORE MORE MORE F2 [TanLn(] 2nd ALPHA (alpha) 0 (y) 1 , 7.24).

Press ENTER and the tangent line is drawn at the high point of the parabola. Compare this line with the zoomed-in view of the curve at point B (in the previous section).

Note that this method of drawing the tangent does not show the slope of the tangent line.

Choose the method you prefer from the two methods given above and draw the tangent at point C where $x = 9$.

(Compare your graph with the one shown in Figure 3.9 in Section 3.2 of the text.)

3.2.2 VISUALIZING THE LIMITING PROCESS This section of the *Guide* is optional, but it might help you understand what it means for the tangent line to be the limiting position of secant lines. Program SECTAN is used to view secant lines between a point $(a, f(a))$ and close points on a curve $y = f(x)$. The program also draws the tangent line at the point $(a, f(a))$. We illustrate with the graph of the function T that was used in the previous two sections, but you can use this program with any graph. (Program SECTAN is in the *TI-86 Program Appendix*.)

Caution: Before using program SECTAN, a function (using x as the input variable) must be entered in the y1 location of the y(x)= list. In order to properly view the secant lines and the tangent line, you must first draw a graph clearly showing the function, the point of tangency (which should be near the center of the graph), and a large enough window so that the close points on either side of the point of tangency can be viewed.

Have $T(x) = -0.804x^2 + 11.644x + 38.114$ in y1. Next, press F2 [WIND] and set xMin = 0, xMax = 7, yMin = 30, and yMax = 90. Press F5 [GRAPH]. Press PRGM F1 [NAMES] and the F-key corresponding to the location of program SECTAN. (Your program list may not look like the one shown on the right.)

Press ENTER to start the program. If you forgot to enter the function in y1 or to draw its graph, press F2 [NO]. Otherwise, press F1 [YES]. At the next prompt, type the input value at the point of tangency. (For this illustration, choose $x = 4$.)	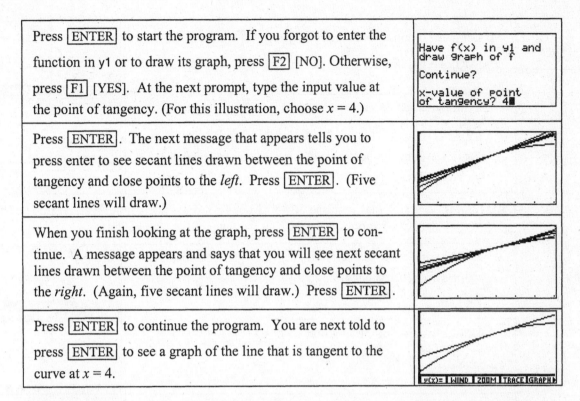
Press ENTER. The next message that appears tells you to press enter to see secant lines drawn between the point of tangency and close points to the *left*. Press ENTER. (Five secant lines will draw.)	
When you finish looking at the graph, press ENTER to continue. A message appears and says that you will see next secant lines drawn between the point of tangency and close points to the *right*. (Again, five secant lines will draw.) Press ENTER.	
Press ENTER to continue the program. You are next told to press ENTER to see a graph of the line that is tangent to the curve at $x = 4$.	

As you watch the graphs, you should notice that the secant lines are becoming closer and closer to the tangent line as the close point moves closer and closer to the point of tangency.

3.2.3 TANGENT LINES AND INSTANTANEOUS RATES OF CHANGE
Sometimes the TI-86 gives results that are not the same as the mathematical results you expect. This does not mean that the calculator is incorrect – it does, however, mean that the calculator programming is using a different formula or definition than the one that you are using. You need to know when the TI-86 produces a different result from what you expect. The results the TI-86 gives involving instantaneous rates of change may vary depending on which derivative is chosen in the MODE screen. For the following discussion, we assume that you have dxDer1 chosen in the last row of the MODE screen as indicated in Figure 1 on page B-1.

In this section, we investigate what the calculator does if you ask it to draw a tangent line where the line cannot be drawn. Consider these special cases:

1. What happens if the tangent line is vertical? We consider the function $f(x) = (x + 1)^{1/3}$ which has a vertical tangent at $x = {}^-1$.

2. How does the calculator respond when the instantaneous rate of change at a point does not exist? We illustrate with $g(x) = |x| - 1$, a function that has a sharp point at $(0, {}^-1)$.

3. Does the calculator draw the tangent line at the break point(s) of a piecewise continuous function? We consider two situations:
   a. $h(x)$, a piecewise continuous function that is continuous at all points, and
   b. $m(x)$, a piecewise continuous function that is not continuous at $x = 1$.

1. Enter the function $f(x) = (x + 1)^{1/3}$ in the y1 location of the y(x)= list. Remember that anytime there is more than one symbol in an exponent and you are not sure of the TI-86's order of operations, enclose the power in parentheses.	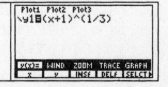

Draw the graph of the function with [EXIT] [F3] [ZOOM] [MORE] [F4] [ZDECM].	
With the graph on the screen, press [MORE] [F1] [MATH] [MORE] [MORE] [F1] [TANLN] [(-)] 1 [ENTER]. Instead of the tangent line being drawn at $x = ^-1$, you see an error message. Press [F5] [QUIT].	ERROR 04 DOMAIN GOTO QUIT This is incorrect!

CAUTION: A vertical tangent line at $x = ^-1$ does not draw (as it should). The slope of the tangent line does not exist, but the tangent line can be drawn.

2.	Clear y1 and enter the function $g(x) = |x| - 1$. The absolute value symbol is typed with [2nd] [X] (MATH) [F1] [NUM] [F5] [abs].	Plot1 Plot2 Plot3 \y1▤abs x-1 x y INSf DELf SELCT round iPart fPart int abs ▶
	Draw the graph of the function with [EXIT] [F3] [ZOOM] [MORE] [F4] [ZDECM].	
	With the graph on the screen, press [MORE] [F1] [MATH] [MORE] [MORE] [F1] [TANLN] 0 [ENTER]. You see an error message. Press [F5] [QUIT].	ERROR 04 DOMAIN GOTO QUIT

This is correct! There is a sharp point on the graph of g at $(0, ^-1)$, and the limiting positions of secant lines from the left and the right at that point are different. A tangent line cannot be drawn on the graph of g at $(0, ^-1)$ according to our definition of instantaneous rate of change.

3a.	Clear y1 and y2 and enter, as indicated, the function $h(x) = \begin{cases} x^2 & \text{when } x \le 1 \\ x & \text{when } x > 1 \end{cases}$. (If you prefer, h can be entered as a single statement in y3, as indicated in Section 1.2.3 of this *Guide*.) Recall that the inequality symbol menu is accessed with [2nd] (TEST).	Plot1 Plot2 Plot3 \y1▤(x²)(x≤1) \y2▤(x)(x>1) y(x)= WIND ZOOM TRACE GRAPH ALL+ ALL- STYLE

Set each part of the function to draw in DOT mode by pressing [MORE] [F3] [STYLE] and continuing to press [F3] until the slanted line turns to a dotted line.

Draw the graph of the function with [EXIT] [F3] [ZOOM] [MORE] [F4] [ZDECM].	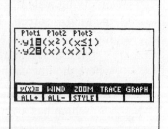
With the graph on the screen, press [MORE] [F1] [MATH] [MORE] [MORE] [F1] [TANLN] 1 [ENTER]. You see an error message. Press [F5] [QUIT].	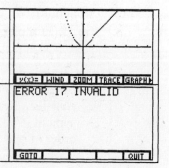 ERROR 17 INVALID GOTO QUIT

This is correct! Even though h is continuous for all values of x, there is no tangent line to the graph of h at the break point $x = 1$. Secant lines drawn using close points on the right and on the left of $x = 1$ do not approach the same slope, so the instantaneous rate of change of h does not exist at $x = 1$.

NOTE: There is <u>no tangent line at the break point</u> of a piecewise continuous function (even if that function is continuous at the break point) unless secant lines drawn through close points to the left and right of that point approach the same value.

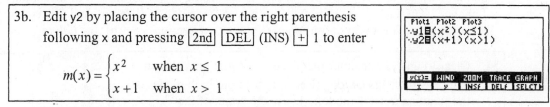

3b. Edit y2 by placing the cursor over the right parenthesis following x and pressing [2nd] [DEL] (INS) [+] 1 to enter

$$m(x) = \begin{cases} x^2 & \text{when } x \le 1 \\ x+1 & \text{when } x > 1 \end{cases}$$

The TI-86 should still be in DOT mode from the previous graph.

Draw the graph of the function with [EXIT] [F3] [ZOOM] [MORE] [F4] [ZDECM].

With the graph on the screen, press [MORE] [F1] [MATH] [MORE] [MORE] [F1] [TANLN] 1 [ENTER]. You see an error message. Press [F5] [QUIT].

This is correct! Because $m(x)$ is not continuous at $x = 1$, the instantaneous rate of change does not exist at that point. The tangent line cannot be drawn on the graph of m when $x = 1$.

The two methods that the TI-86 uses to compute instantaneous rates of change give the same numerical results for smooth continuous functions. However, the results for the two methods are not the same for all functions. The method the TI-86 uses to compute instantaneous rates of change with each of the two methods is explained in detail in Section 4.3.1 of this *Guide*. We now see what results are given for the previous investigations when the calculator's mode screen is set to dxNDer instead of dxDer1.

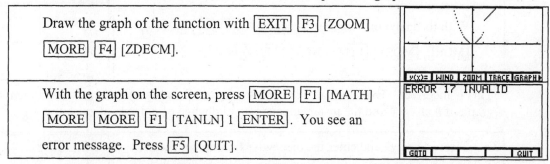

1. Press [2nd] [MORE] (MODE) and choose dxNDer in the last row of the MODE screen. Enter the function $f(x) = (x + 1)^{1/3}$ and draw its graph with [EXIT] [F3] [ZOOM] [MORE] [F4] [ZDECM].

With the graph on the screen, press [MORE] [F1] [MATH] [MORE] [MORE] [F1] [TANLN] [(-)] 1 [ENTER].

CAUTION: Note that the TI-86 correctly draws the vertical tangent. However, the TI-86 says the slope is 100, but the slope of this line at $(^-1, 0)$ does not exist because the tangent is vertical.

2. Next, clear y1 and enter the function $g(x) = |x| - 1$. (Type the absolute value symbol with [2nd] [X] (MATH) [F1] [NUM] [F5] [abs].) Draw the graph with ZDECM.

With the graph on the screen, press [MORE] [F1] [MATH] [MORE] [MORE] [F1] [TANLN] 0 [ENTER].

This is incorrect! There is a sharp point on the graph of g at $(0, ^-1)$, and the limiting positions of secant lines from the left and the right at that point are different. A tangent line cannot be drawn on the graph of g at $(0, ^-1)$ according to *our* definition of instantaneous rate of change. The TI-86's definition using the dxNDer mode is different, and this is why the line is drawn. (The TI-86 definition is explained in Section 3.5.1 of this *Guide*.)

3a.   Clear y1 and y2 and enter the piecewise function h shown page B–44 of this *Guide*. (Remember to put both pieces of the function in DOT mode.) Draw the graph of the function with ZDECM.

With the graph on the screen, press [MORE] [F1] [MATH] [MORE] [MORE] [F1] [TANLN] 1 [ENTER].

This is incorrect! Even though h is continuous for all values of x, there is no tangent to the graph of h at $x = 1$ and the instantaneous rate of change of h does not exist at $x = 1$.

3b.   Clear y1 and y2 and enter the piecewise function m shown page B–45 of this *Guide*. (Remember to put both pieces of the function in DOT mode.) Draw the graph of the function with ZDECM.

With the graph on the screen, press [MORE] [F1] [MATH] [MORE] [MORE] [F1] [TANLN] 1 [ENTER].

This is incorrect! Because m is not continuous at $x = 1$, the instantaneous rate of change does not exist at that point. Press [2nd] [MORE] (MODE) and choose dxDer1.

CAUTION: Be certain that the instantaneous rate of change exists at a point and the tangent line can be drawn at that point before using your calculator to draw a tangent line or use the slope of the tangent line that may be printed on the screen.

 ## 3.3 Derivatives

There are no new calculator techniques in this section, but we illustrate a new calculation.

3.3.1 CALCULATING PERCENTAGE RATE OF CHANGE   Suppose the growth rate of a population is 50,000 people per year and the current population size is 200,000 people.

What is the percentage rate of change of the population? The answer is 25% per year.	50000/200000 .25 Ans*100 25
Suppose instead that the current population size is 2 million. What is the percentage rate of change? The answer is 2.5% per year, which is a much smaller percentage rate of change.	50000/(2*10^6) .025 Ans*100

3.4 Numerically Finding Slopes

Using your calculator to find slopes of tangent lines does not involve a new procedure. However, the techniques that are discussed in this section allow you to repeatedly apply a method of finding slopes that gives quick and accurate results.

3.4.1a NUMERICALLY ESTIMATING SLOPES ON THE HOME SCREEN   Finding the slopes of secant lines joining the point at which the tangent line is drawn to increasingly close points on a function to the left and right of the point of tangency is easily done using your TI-86. Suppose we want to numerically estimate the slope of the tangent line at $t = 8$ to the graph of the function that gives the number of polio cases in 1949: $y = \dfrac{42{,}183.911}{1 + 21{,}484.253e^{-1.248911t}}$

where $t = 1$ on January 31, 1949, $t = 2$ on February 28, 1949, etc. (See Example 1 in the text.)

Enter the polio cases equation in the y1 location of the y(x)= list. (Carefully check your entry of the equation, and be sure to use parentheses around the denominator and the exponent.) We now evaluate the slopes of secant lines that join close points to the left of $x = 8$ with $x = 8$.	Plot1 Plot2 Plot3 \y1▯...3e^(-1.248911x)) y(x)= WIND ZOOM TRACE GRAPH x y INSf DELf SELCT▶
On the home screen, type in the expression shown to the right to compute the slope of the secant line joining the close point where $x = 7.9$ and the point of tangency where $x = 8$.	(y1(8)-y1(7.9))/(8-7.9) 13159.6827248

Record on paper each slope, to at least 1 more decimal place than the desired accuracy, as it is computed. You are asked to find the nearest whole number that these slopes are approaching, so record at least one decimal place in your table of slopes.

Press [2nd] [ENTER] (ENTRY) to recall the last entry, and then use the arrow keys to move the cursor over the 9 in the "7.9". Press [2nd] [DEL] (INS) and press 9 to insert another 9 in <u>both</u> places that 7.9 appears. Press [ENTER] to find the slope of the secant line joining $x = 7.99$ and $x = 8$.	(y1(8)-y1(7.9))/(8-7.9)
	        13159.6827248
	(y1(8)-y1(7.99))/(8-7.99)
	        13170.6176627

Continue in this manner, recording each result on paper, until you can determine to which value the slopes from the left seem to be getting closer and closer.
 	
It appears that the slopes of the secant lines from the left are approaching 13,170 cases per month.
	13170.187131
	(y1(8)-y1(7.9999))/(8-7.9999)
	        13170.12882
	(y1(8)-y1(7.99999))/(8-7.99999)
	        13170.1229

We now evaluate the slopes joining the point of tangency and nearby close points to the *right* of $x = 8$. Clear the screen, recall the last expression with [2nd] [ENTER] (ENTRY), and edit it so that the nearby point is $x = 8.1$. Press [ENTER] to calculate the secant line slope.	(y1(8)-y1(8.1))/(8-8.1) 13146.3842089

Continue in this manner as you did when calculating slopes to the left, but each time insert a 0 before the "1" in two places in the close point. Record each result on paper until you can determine the value the slopes from the right are approaching.

```
                13170.053799
(y1(8)-y1(8.0001))/(8
-8.0001)
                 13170.11549
(y1(8)-y1(8.00001))/(
8-8.00001)
                  13170.1214
```

It appears that the slopes of the secant lines from the right are approaching 13,170 cases per month.

When the slopes from the left and the slopes from the right approach the same number, that number is the slope of the tangent line at the point of tangency. In this case, we estimate the slope of the tangent line to be 13,170 cases per month.

3.4.1b NUMERICALLY ESTIMATING SLOPES USING THE TABLE   The process discussed

in Section 3.4.1a of this *Guide* can be done in fewer steps and with fewer keystrokes when you use the TI-86 TABLE. The point of tangency is $x = 8$, $y = $ y1(8), and let's call the close

point $(x, $ y1$(x))$. Then, slope $= \dfrac{\text{rise}}{\text{run}} = \dfrac{y1(x) - y1(8)}{x - 8}$. We illustrate numerically estimating the

slope using the TABLE with the logistic function given in Section 3.4.1a of this *Guide*.

Have the polio cases equation given in the previous section in the y1 location of the y(x)= list. Enter the above slope formula in a different other location, say y2. (Remember to enclose both numerator and denominator of the slope formula in parentheses.) Turn off y1 because we are only considering the output from y2.	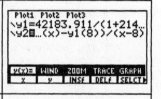

Press TABLE F2 [TBLST] and choose ASK in the Indpnt: location. (See page B-8 for more specific instructions.)

Access the table with F1 [TABLE] and delete or type over any previous entries in the x column. Enter values for x, the input of the close point, so that x gets closer and closer to 8 from the left.

```
  x     | y2
7.9     | 13159.68
7.99    | 13170.62
7.999   | 13170.19
7.9999  | 13170.13
7.99999 | 13170.12

x=7.99999
TBLST SELCT  x    y
```

• Although it did not happen in this illustration, the calculator might switch the input values you enter to scientific notation and display rounded output values so that the numbers can fit on the screen in the space allotted for outputs of the table. If this happens, you should position the cursor over each output value and record on paper as many decimal places as necessary in order to determine the limit to the desired degree of accuracy.

Repeat the process, entering values for x, the input of the close point, so that x gets closer and closer to 8 from the right.

Remember to highlight the output values and record on paper the slope to at least one more decimal position than the desired accuracy.

```
  x    | y2
8.1    | 13146.38
8.01   | 13169.28
8.001  | 13170.05
8.0001 | 13170.12
8.00001| 13170.12

y2=13170.1214
TBLST SELCT  x    y
```

The limit is estimated to be the value that the limits from the left and the right are approaching: 13,170 cases per month.

NOTE: You may wish to leave the slope formula in y2 as long as you need it. Remember to change the input value at the point of tangency the next time you use this formula. Turn y2 off when you are not using it with F5 [SELCT].

3.5 Algebraically Finding Slopes

The TI-86 does not find algebraic formulas for slope, but you can use the built-in numerical derivatives and draw the graph of a derivative to check any formula that you find algebraically.

3.5.1 UNDERSTANDING YOUR CALCULATOR'S SLOPE FUNCTION The TI-86 has two derivative functions: nDer (used in dxNDer mode) and der1 (used in dxDer1 mode). The numerical derivative nDer uses the slope of a secant line to approximate the slope of the tangent line at a point on the graph of a function. However, instead of using a secant line through the point of tangency and a close point, nDer uses the slope of a secant line through two close points that are equally spaced from the point of tangency.

The secant line joining the points $(a-k, f(a-k))$ and $(a+k, f(a+k))$ and the line tangent to the graph of f where $x = a$ are shown in Figure 4. Notice that the slopes of the secant line and tangent appear to be close to the same value even though these are different lines.

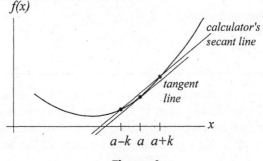

Figure 4

As k gets closer and closer to 0, the two points through which the secant line passes move closer and closer to a. Provided the slope of the tangent line exists, the limiting position of the secant line will be the tangent line.

The calculator's notation for the slope of the secant line shown in Figure 4 is

$$\text{nDer(function, symbol for input variable, } a)$$

The k that determines the locations of the points on the secant line can be set to different values, but specifying this value is optional. If it is not given, the calculator automatically uses $k = 0.001$ (or whatever the current setting may be). The TI-86 uses the symbol δ for k, and it also has a value called *tol*. This is the TI-86 tolerance – a quantity that can affect the speed and accuracy of some calculations.

The TI-86 also computes for many functions the slope of the tangent line. This derivative, der1, computes for many built-in functions the exact derivative. Even though der1 is more accurate than nDer, it is more restrictive because it does not work for all functions.

Any smooth, continuous function will do for our explorations, so let's investigate and compare nDer and der1 with the function in Example 3 of Section 3.5 of *Calculus Concepts*: $f(x) = 2\sqrt{x}$. First we find nDer, the slope of the secant line between the two points $(0, f(0))$ and $(2, f(2))$. That is, we are finding the slope of the line between the points $(a-k, f(a-k))$ and $(a+k, f(a+k))$ for $a = 1$ and $k = 1$. Enter $f(x) = 2\sqrt{x}$ in y1.

Set the value of k by pressing ⟨2nd⟩ 3 (MEM) ⟨F4⟩ (TOL) and enter $k = \delta =$ 1. For our purposes, have tol = 10^{-6} = 1E-6.	Ploti Plot2 Plot3 \y1■2√x Y(X)= WIND ZOOM TRACE GRAPH x y INSf DELf SELCT▶	TOLERANCE tol=1E-6 δ=1

On the home screen, access nDer(with ⎡2nd⎤ ⎡÷⎤ (CALC) ⎡F2⎤ [nDer], and enter nDer(y1, x, 1). Access der1(with ⎡F3⎤ [der1] and enter der1(y1, x, 1).		nDer(y1,x,1)
		                1.41421356237
		der1(y1,x,1)
		                            1
		
		eva1F nDer der1 der2 fnInt ▶

Press ⎡2nd⎤ 3 (MEM) ⎡F4⎤ (TOL) and change the value of k from 1 to 0.1. On the home screen, recall the last two entries using ⎡2nd⎤ ⎡ENTER⎤ (ENTRY).	TOLERANCE
	tol=1ᴇ⁻6
	δ=.1
		1.41421356237
		der1(y1,x,1)
		                            1
		nDer(y1,x,1)
		                1.0012555012
		der1(y1,x,1)
		                            1

Repeat the process for $k = 0.001$ and $k = 0.0001$. Note how the results are changing.	TOLERANCE
	tol=1ᴇ⁻6
	δ=.001
		1.0012555012
		der1(y1,x,1)
		                            1
		nDer(y1,x,1)
		                1.00000012505
		der1(y1,x,1)
		                            1

As k becomes smaller and smaller, the secant line slope is becoming closer and closer to 1.	TOLERANCE
	tol=1ᴇ⁻6
	δ=.0001
		1.00000012505
		der1(y1,x,1)
		                            1
		nDer(y1,x,1)
		                1.0000000015
		der1(y1,x,1)
		                            1

- When we observe the secant line slopes, the nDer values, we reach the conclusion that the limiting value of the slopes is 1. However, this is the result of a numerical investigation, not an algebraic proof. Note that the value of der1 remains constant at 1 no matter what k is used. This is because der1 computes the tangent line slope exactly for this function.

In the table below, the first row lists some values of a, the input of a point of tangency, and the second row gives the actual slope (to 7 decimal places) of the tangent line at those values. The algebraic method gives the exact slope of the line tangent to the graph of f at these input values.

Use your calculator to verify the values in the third through sixth rows that give the values of der1 and nDer (to 7 decimal places), the slope of the secant line between $(a - k, f(a - k))$ and $(a + k, f(a + k))$ for the indicated values of k. Before calculating each secant line slope, set the value of $k = \delta$ on the TOLERANCE screen accessed with ⎡2nd⎤ 3 (MEM) ⎡F4⎤ (TOL).

a = input of point of tangency	2.3	5	12.82	62.7
slope of tangent line = $f'(a)$	0.6593805	0.4472136	0.2792904	0.1262892
value of der1(y1, x, a)	0.6593805	0.4472136	0.2792904	0.1262892
slope of secant line, $k = 0.1$	0.6595364	0.4472360	0.2792925	0.1262892
slope of secant line, $k = 0.01$	0.6593820	0.4472138	0.2792904	0.1262892
slope of secant line, $k = 0.001$	0.6593805	0.4472136	0.2792904	0.1262892
slope of secant line, $k = 0.0001$	0.6593805	0.4472136	0.2792904	0.1262892

You can see that der1(y1, x, a) gives the slope of the tangent line and that the secant line slope is very close to the slope of the tangent line for small values of k. For this function, the secant line slope does a great job of approximating the slope of the tangent line when k is very small.

Will the slope of this secant line always do a good job of approximating the slope of the tangent line when $k = 0.001$? Yes, it generally does, as long as the instantaneous rate of change exists at the input value at which you evaluate nDer. Therefore, from this point forward we use $k = 0.001$ whenever we evaluate nDer. Return to the TOLERANCE screen and set $k = \delta = 0.001$.

Note that when the instantaneous rate of change does not exist, neither nDer nor der1 should be used to approximate something that does not have a value!

You will benefit from reading again Section 3.2.3 of this *Guide*, which illustrates several cases when the TI-86's slope function nDer gives a value for the slope when it does not exist. For instance, consider $f(x) = |x|$ for which the instantaneous rate of change does not exist at $x = 0$ because the graph of f has a sharp point there. Note that you can enter the function in the y(x)= list or enter it directly into the derivative instruction on the home screen.

| Find the TI-86's derivatives at $x = 0$ for $f(x) = |x|$. Note that the slope of the secant line is 0, but the instantaneous rate of change does not exist at $x = 0$. | 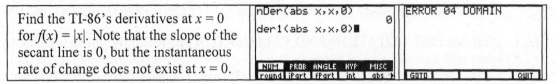 |

- When the der1 instruction is entered, it gives an error message (as it should). Whenever you receive an error message, you need to carefully examine what you are doing and see if there is a mathematical reason for the error message (as is the case here).

We recommend that you keep the TI-86's MODE to those settings given in Figure 1 on page B-1. Check that your calculator has those settings before proceeding. Because it is more mathematically compatible with our definitions, we also suggest that you use der1 rather than nDer whenever possible.

Chapter 4   Determining Change: Derivatives

4.3 More Simple Rate-of-Change Formulas

The TI-86 only approximates or calculates numerical values of slopes – it does not give the slope in formula form. You also need to remember that nDer computes slope by a different definition than we use and that you may not obtain the correct answer when using nDer with functions at points where the slope does not exist. For this reason we use der1, not nDer.

4.3.1 DERIVATIVE NOTATION AND CALCULATOR NOTATION
In addition to learning when your calculator gives an acceptable answer for a derivative and when it does not, you also need to understand the differences and similarities in mathematical derivative notation and calculator notation. The notation that we use for the calculator's numerical derivative is der1($f(x)$, x, x). The correspondence between the mathematical derivative notation $\frac{df(x)}{dx}$ and the calculator's notation der1($f(x)$, x, x) is illustrated in Figure 5.

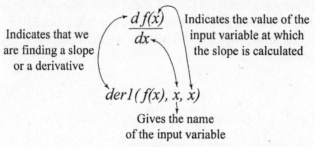

Figure 5

We illustrate another use of the TI-86's derivative by constructing Table 4.7 in Section 4.3 of *Calculus Concepts*. The table lists x, $y = f(x) = e^x$, and $y' = \frac{df}{dx}$ for 4 different inputs. We next evaluate this function at these 4 and several other inputs.

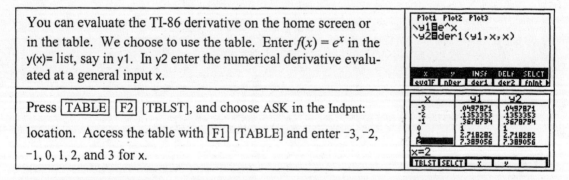

You can evaluate the TI-86 derivative on the home screen or in the table. We choose to use the table. Enter $f(x) = e^x$ in the y(x)= list, say in y1. In y2 enter the numerical derivative evaluated at a general input x.
Press TABLE F2 [TBLST], and choose ASK in the Indpnt: location. Access the table with F1 [TABLE] and enter −3, −2, −1, 0, 1, 2, and 3 for x.

It appears that the derivative values are the same as the function outputs. In fact, this is a true statement for all inputs of f – this function is its own derivative!

You can use the methods discussed in Section 1.3.2a of this Guide to find the values used in Table 4.9 to numerically estimate $\lim\limits_{h \to 0} \frac{2^h - 1}{h}$. Instead of this, we explore an alternate method of confirming that $\frac{d(2^x)}{dx} = (\ln 2)\, 2^x$.

Press $\boxed{\text{GRAPH}}$ $\boxed{\text{F1}}$ [y(x)=] and edit y1 to be $g(x) = 2^x$. Access the statistical lists and clear any previous entries from L1, L2, L3, and L4. Enter the x-values shown above in L1. Highlight L2 and type y1(L1).	
Press $\boxed{\text{ENTER}}$ to fill L2 with the function outputs. Then, highlight L3 and type y2(L1).	
Press $\boxed{\text{ENTER}}$ to place in list L3 the derivative of y1 evaluated at the inputs in L1. Note that these values are not the same as the function outputs. To see what relation the slopes have to the function outputs, press $\boxed{\blacktriangleright}$ and highlight L4. Enter L3 $\boxed{\div}$ L2.	

It appears that the slope values are a multiple of the function output. In fact, that multiple is $\ln 2 \approx 0.693147$. Thus we confirm this slope formula: If $g(x) = 2^x$, then $\dfrac{dg}{dx} = (\ln 2)\,2^x$.

4.3.2a CALCULATING $\dfrac{dy}{dx}$ AT SPECIFIC INPUT VALUES

The previous two sections of this *Guide* examined the calculator's numerical derivatives, nDer($f(x)$, x, a) and der1($f(x)$, x, a), and illustrated that they equal or give a good approximation to the slope of the tangent line at points where the instantaneous rate of change exists. You can also evaluate the calculator's derivatives from the graphics screen using the CALC menu. Whichever derivative you have selected in the MODE screen is the one that is used, and it is called *dy/dx*. We illustrate this use with the function in part *a* of Example 2 in Section 4.3 of *Calculus Concepts*.

Clear all previously-entered functions in the y(x)= list. Enter $f(x) = 12.36 + 6.2 \ln x$ in y1. We want to draw a graph of *f*. Realize that $x \geq 0$ because of the log term. Choose some value for xMax, say 5. Use $\boxed{\text{F3}}$ [ZOOM] $\boxed{\text{MORE}}$ $\boxed{\text{F1}}$ [ZFIT] to set the height of the graph.	
With the graph on the screen, press $\boxed{\text{MORE}}$ $\boxed{\text{F1}}$ [MATH] $\boxed{\text{F2}}$ [dy/dx]. Use $\boxed{\blacktriangleleft}$ or $\boxed{\blacktriangleright}$ to move to some point on the graph.	
Press $\boxed{\text{ENTER}}$ and the slope of the function is calculated at the input you chose in the previous step.	
To find the derivative evaluated at a specific value of x, you can type in the desired input instead of pressing the arrow keys. Press $\boxed{\text{CLEAR}}$ $\boxed{\text{EXIT}}$ (to return the menu) and $\boxed{\text{F1}}$ [MATH] $\boxed{\text{F2}}$ [dy/dx] 3 $\boxed{\text{ENTER}}$.	

The slope $dy/dx \approx 2.066667$ appears at the bottom of the screen. Return to the home screen and press [x-VAR]. The TI-86's memory location for x has been updated to 3. Now type both numerical derivative instructions (evaluated at 3) as shown to the right. This dy/dx value you saw on the graphics screen is the der1 value because we have the MODE set to dxDer1.

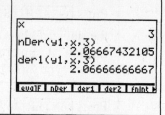

You can use the ideas presented above to check your algebraic formula for the derivative. We next investigate this procedure.

4.3.2b NUMERICALLY CHECKING SLOPE FORMULAS It is always a good idea to check your answer. Although your calculator cannot give you an algebraic formula for the derivative function, you can use numerical techniques to check your algebraic derivative formula. The basic idea of the checking process is that if you evaluate your derivative and the TI-86's derivative at several randomly chosen values of the input variable and the outputs are basically the same values, your derivative is *probably* correct.

These same procedures are applicable when you check your results (in the next several sections) after applying the Sum Rule, the Chain Rule, or the Product Rule. We use the function in part *c* of Example 2 in Section 4.3 of *Calculus Concepts* to illustrate.

Enter $m(r) = \frac{8}{r} - 12\sqrt{r}$ in y1 (using x as the input variable).

Compute $m'(r)$ using pencil and paper and the derivative rules. Enter this function in y2. (What you enter in y2 may or may not be the same as what appears to the right.)

Enter the TI-86's derivative of y1 (evaluated at a general input x) in y3. Because you are interested in seeing if the outputs of y2 and y3 are the same, turn off y1.

Press [TABLE] [F2] [TBLST] and choose ASK in the Indpnt: location. Access the table with [F1] [TABLE] and enter at least three different values for x.

The table gives strong evidence that that y2 and y3 are the same function.

4.3.2c GRAPHICALLY CHECKING SLOPE FORMULAS When it is used correctly, a graphical check of your algebraic formula works well because you can look at many more inputs when drawing a graph than when viewing specific inputs in a table. We illustrate this use with the function in part *d* of Example 2 in Section 4.3 of *Calculus Concepts*.

Enter $j(y) = 17{,}000\left(1 + \frac{0.025}{12}\right)^{12y}$ in y1, using x as the input variable. Next, using pencil and paper and the derivative rules, compute $j'(y)$. Enter this function in y2.

Enter the TI-86's numerical derivative of y1 (evaluated at a general input x) in y3. Before proceeding, turn off the graphs of y1 and y3.

NOTE: y1 = 17000(1 + 0.025/12)^(12x) and we entered y2 = 17000*ln(1.025288)(1.025288^x).

To graphically check your derivative formula, you now need to find a good graph of y2. This function is not in a context with a given input, so the time it takes to draw a graph is shortened if you know the approximate shape of the graph. Note that the graph of the function in y2 is an increasing exponential curve.

Start with F3 [ZOOM] MORE F4 [ZDECM] or F3 [ZOOM] F4 [ZSTD]. Neither of these shows a graph (because of the large coefficient in *j*), but you can press TRACE to see some of the output values. Use those values to reset the window.	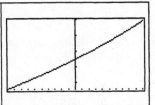

- The graph shown above was drawn in the window [−10, 10] by [330, 550], but any view that shows a graph will do.

Now, press F1 [y(x)=], turn off y2, and turn on y3. (Recall that y3 holds the formula for the derivative of y1 as computed numerically by the TI-86.) Press F5 [GRAPH]. Note that you are drawing the graph of y3 in *exactly the same window* in which you graphed y2.	

If you see the same graph, your algebraic formula (in y2) is very likely correct.

 ## 4.4 The Chain Rule

You probably noticed that checking your answer for a slope formula graphically is more difficult than checking your answer using the TI-86 table if you have to spend a lot of time finding a window in which to view the graph. Practice with these methods will help you determine which is the best to use to check your answer. (These ideas also apply to Section 4.5.)

4.4.1 SUMMARY OF CHECKING METHODS   Before you begin checking your answer, make sure that you have correctly entered the function. It is very frustrating to miss the answer to a problem because you have made an error in entering a function in your calculator. We summarize the methods of checking your algebraic answer using the function in Example 3 of Section 4.4 of *Calculus Concepts*.

Enter the function $P(t) = \dfrac{84.4}{1+33.6\,e^{-0.484t}}$ in the y1 location, your answer for P' in y2, and the TI-86 derivative in y3. Turn off y1. Note that your entry in y2 may be different, but we entered y2 = 1372.546e^(−0.484x)/(1+33.6e^(−0.484x))^2.	
Go to the table, which has been set to ASK mode. Enter at least 3 input values. (Because this problem is in a context, read the problem to see which inputs make sense.) It seems that the answer in y2 is probably correct!	
If you prefer a graphical check, the problem states that the equation is valid between 1980 and 2001 (and the input is the number of years after 1980). So, turn off y3, set xMin = 0, xMax = 21, and draw a graph of y2 using ZFIT. Then, using the same window, draw a graph of y3 with y2 turned off. The graphs are the same, again suggesting a correct answer in y2.	

Chapter 5   Analyzing Change:
Applications of Derivatives

 5.2 Relative and Absolute Extreme Points

Your calculator can be very helpful for checking your analytic work when you find optimal
points and points of inflection. When you are not required to show work using derivative
formulas or when an approximation to the exact answer is all that is required, it is a simple
process to use your calculator to find optimal points and inflection points.

5.2.1a FINDING X-INTERCEPTS OF SLOPE GRAPHS   Where the graph of a function has a
relative maximum or minimum, the slope graph has a horizontal tangent. Where the tangent
line is horizontal, the derivative of the function is zero. Thus, finding where the slope graph
crosses the input axis is the same as finding the input of a relative extreme point.

Consider, for example, the model for Acme Cable Company's revenue for the 26 weeks
after it began a sales campaign, where x is the number of weeks since Acme began sales:

$$R(x) = -3x^4 + 160x^3 - 3000x^2 + 24,000x \text{ dollars}$$

In Example 2 of Section 5.2 of *Calculus Concepts,* we are first asked to determine when
Acme's revenue peaked during the 26-week interval.

Enter R in the y1 location of the y(x)= list. Enter either the TI-86's derivative or your derivative in the y2 location. Turn off y1. (If you use your derivative, be sure to use one of the methods at the end of Chapter 4 in this *Guide* to check that your derivative and the TI-86 derivative are the same.)	Plot1 Plot2 Plot3 \y1=-3x^4+160x^3-300... \y2■der1(y1,x,x) y(x)= WIND ZOOM TRACE GRAPH x y INSf DELf SELCT▶
The statement of the problem indicates that x should be graphed between 1 and 26. Set this horizontal view, and draw the slope graph in y2 with [F3] [ZOOM] [MORE] [F1] [ZFIT]. Reset the window to a better view for this illustration: yMin = ‑800 and yMax = 3000. Redraw the slope graph with [F5] [GRAPH].	
With the graph on the screen, find the intercepts of the slope graph with [MORE] [F1] [MATH] [F1] [ROOT]. Press and hold [▶] to move the cursor near to, but still to the *left* of, the rightmost x-intercept.	Left Bound? X=19.055555556 y=929.36831273
Press [ENTER] to mark the location of the left bound. Use [▶] to move the cursor near to, but to the *right* of, the rightmost x-intercept.	Right Bound? X=20.246031746 y=-309.9443964
Press [ENTER] to mark the location of the right bound. You are next asked to provide a guess. Any value near the intercept will do. Use [◀] to move the cursor near the intercept.	Guess? X=20.047619048 y=-57.68837056

- Note that the calculator has marked (at the top of the screen) the left and right bounds
  with small triangles. The x-intercept must lie between these two bound marks. If you
  incorrectly mark the interval, you may not get an answer.

Press ENTER. The location of the x-intercept is displayed. We see that $R'(x) = 0$ at $x = 20$.

(The numerical process used the find the root sometimes includes small rounding errors. The output (y) at the root should be 0 but is printed on the screen as ⁻0.000000005.)

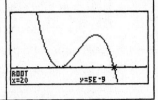

As requested in part *b* of Example 2, we now need to determine if the slope graph crosses the x-axis, only touches the x-axis, or does neither at the other location that is an intercept. Note that you can find that $x = 10$ at this point using the same procedure as described above.

Zoom in with ZIN (see Section 1.3.2b) or BOX (see Section 3.2.1a) as many times as necessary to magnify the portion of the graph around $x = 10$ in order to examine it more closely. We choose to use BOX, but both work equally well.	
After magnifying the graph several times, we see that the graph just touches and does not cross the x-axis near $x = 10$. (Press F4 [TRACE] and trace as near as possible to $x = 10$.) We see that $x = 10$ does not yield an extreme point on the graph of R.	
We are asked to find the absolute maximum, and we know that it occurs at one of the endpoints of the interval or at a zero of the slope graph. So, return to the home screen and find the outputs of R at the endpoints of the interval ($x = 0$ and $x = 26$) as well as the output at the "crossing" root of the slope graph ($x = 20$).	

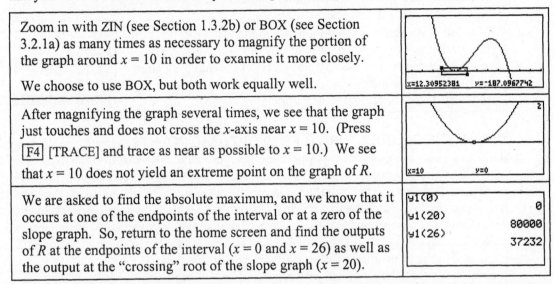

We see that Acme's revenue was greatest at 20 weeks after they began the sales campaign.

5.2.1b FINDING ZEROS OF SLOPE FUNCTIONS USING THE SOLVER   You may find it more convenient to use the TI-86 SOLVER rather than the graph and the CALC menu to find the x-intercept(s) of the slope graph. We illustrate using Acme Cable's revenue function.

Enter R in the y1 location of the y(x)= list. Enter either the TI-86's derivative or your derivative in the y2 location. Turn off y2.	
Draw a graph of y1 to obtain a guess as to the location of any relative maxima for use in the SOLVER. We are told that x is between 1 and 26. Set this horizontal view, and draw the graph of y1 with F3 [ZOOM] MORE F1 [ZFIT].	
Reset yMax to a larger value, say 95,000, to better see the high point on the graph. Graph R and press TRACE. Hold down ▶ until you have an estimate of the input location of the high point. The maximum seems to occur when x is near 19.8.	

Remember that we want to solve the equation $R'(x) = 0$ to find where the graph of R has a maximum or minimum. We are using as a guess of the answer the value we obtained by tracing the graph of R.

Access the SOLVER with [2nd] [GRAPH] (SOLVER). Clear what is there from a previous problem, type y2 as the left side of the equation, and press [ENTER]. Enter 0 as the right-hand side of the equation; type the guess $x \approx 19.8$. With the cursor on the line containing x, press [F5] [SOLVE]. The solution $x = 20$ is found.

```
exp=y2
  exp=0
•x=20.000000000003
 bound={-1E99,1E99}
•left-rt=2E-9

GRAPH WIND ZOOM TRACE SOLVE
```

NOTE: Recall that calculators use numerical algorithms to find zeros. You may or may not obtain the exact value 20. Always round the answer obtained from the SOLVER so that it makes sense in the problem context (here, round to a whole number.)

Note that if you enter $x \approx 10$ as a guess in the SOLVER, it returns the solution $x = 10$. You should not rely only on the SOLVER – you must also examine the function graph to see that only 20 is the location of the maximum. Acme's revenue was greatest at 20 weeks after they began the sales campaign.

```
exp=y2
  exp=0
•x=10
 bound={-1E99,1E99}
•left-rt=0

GRAPH WIND ZOOM TRACE SOLVE
```

5.2.1c USING THE CALCULATOR TO FIND OPTIMAL POINTS Once you draw a graph of a function that clearly shows the optimal points, your calculator can find the location of those high points and low points without using calculus. However, we recommend not relying only on this method because your instructor may ask you to show your work using derivatives. If so, this method would probably earn you no credit! This method does give a good check of your answer, and we illustrate it using Acme Cable's revenue function R from Section 5.2.1a.

Enter R in the y1 location of the y(x)= list. The problem indicates that x should be graphed between 1 and 26, so set this horizontal view. Draw the function graph with [F3] [ZOOM] [MORE] [F1] [ZFIT]. Press [F2] [WIND] and reset yMax to 95,000. Press [F5] [GRAPH].	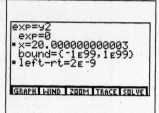
With the graph of R on the screen, press [MORE] [F1] [MATH] [F5] [FMAX]. Use [▶] to move the cursor near, but still to the *left* of, the high point on the curve.	Left Bound? X=18.063492064 y=78288.735429
Press [ENTER] to mark the left bound of the interval. Use [▶] to move the cursor to the *right* of the high point on the curve.	Right Bound? X=21.833333333 y=77456.479167
Press [ENTER] to mark the right bound of the interval. Use [◀] to move the cursor to your guess for the high point on the curve.	Guess? X=20.047619048 y=79998.630802

CAUTION: Notice the small arrowheads at the top of the screen that mark the bounds of the interval. The TI-86 returns the highest point that is within the bounded interval that you have marked when you press [ENTER]. If the relative maximum does not lie in this interval, the TI-86 will return, instead, the highest point in the interval you marked (usually an endpoint).

Press ENTER and the location of the relative maximum (*x*) and the relative maximum value (*y*) are displayed.

(As previously mentioned, you may not see the exact answer due to rounding errors in the numerical routine used by the TI-86. Always round your answer to make sense in the context.)

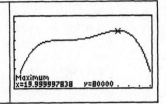

- The method shown in this section also applies to finding the relative minimum values of a function. The only difference is that to find the relative minimum instead of the relative maximum, initially press GRAPH MORE F1 [MATH] F4 [FMIN].

5.3 Inflection Points

As was the case with optimal points, your calculator can be very helpful in checking algebraic work when you find points of inflection. You can also use the methods illustrated in Section 5.2.1c of this *Guide* to find the location of any maximum or minimum points on the graph of the first derivative to find the location of any inflection points for the function.

5.3.1 FINDING *X*-INTERCEPTS OF A SECOND-DERIVATIVE GRAPH
We first look at using the algebraic method of finding inflection points – finding where the graph of the second derivative of a function *crosses* the input axis.

To illustrate, we consider a model for the percentage of students graduating from high school in South Carolina from 1982 through 1990 who entered post-secondary institutions:

$$f(x) = -0.1057x^3 + 1.355x^2 - 3.672x + 50.792 \text{ percent}$$

where *x* is the number of years after 1982.

Enter *f* in the y1 location of the y(x)= list, the first derivative of *f* in y2, and the second derivative of *f* in y3. (Be careful not to round any decimal values.) Turn off y1 and y2.	Plot1 Plot2 Plot3 \y1=-.1057x^3+1.355x... \y2=-.1057(3x²)+1.35... \y3⊟-.1057(6x)+1.355... y(x)≡ WIND ZOOM TRACE GRAPH x y INSf DELf SELCTⴕ
We are given the input interval 1982 through 1990, so $0 \le x \le 8$. Either use ZFIT or choose some appropriate vertical view. We use $-4 \le y \le 4$. Because we are looking for the *x*-intercept(s) of the second derivative graph, any view that shows the line crossing the horizontal axis is okay to use.	y(x)≡ WIND ZOOM TRACE GRAPHⴕ
Use the methods indicated in Section 5.2.1a to find where the second derivative graph crosses the *x*-axis. (Note that when you are asked to give the inflection *point* of *f*, you should answer with both the input and an output of the original function.)	ROOT x=4.2730999685 y=0

Return to the home screen and enter x-VAR. The *x*-value you just found as the *x*-intercept remains stored in the x location until you change it by tracing, using the SOLVER, and so forth. Find the *y*-value by substituting this *x*-value into y1.	x
	        4.27309996846
	y1(Ans)
	        51.5954865334
	y1(x)
	        51.5954865334

At some point, be sure to examine a graph of the function and verify that an inflection point does occur at the point you have found. To do this, turn off y3, turn on y1, and use ZFIT to draw the graph. Trace near where $x \approx 4.27$ and $y \approx 51.6$. The graph of *R* confirms that an inflection point occurs at this point.	x=4.253968254 y=51.554965337

EXPLORE: What would the line tangent to the graph of *f* look like at the inflection point? Use the graph of y1 and the first method explained in Section 3.2.1b of this *Guide* (with *x* = 4.27) to see if you are correct.

EXPLORE: What is true about the graph of *f′* at the inflection point? Use the graph of y2 and the trace cursor to determine if your guess is correct.

- The TI-86 usually draws an accurate graph of the first derivative of a function when you use der1. It also draws the graph of and computes numerically the second derivative of many functions with a command called der2. If you are in doubt about your calculation of the first and second derivative formulas, use the calculator's derivatives to avoid making a mistake in the location of the inflection point.

Enter *f* in the y1 location of the y(x)= list, the first derivative of *f* in y2, and the second derivative of *f* in y3, using the TI-86's derivative for each derivative that you enter. Access der2 with 2nd ÷ (CALC) F4 [der2]. Turn off y1. Be sure that you set xMin = 0 and xMax = 8 (as the problem directions indicate.)	Plot1 Plot2 Plot3 \y1=-.1057x^3+1.355x... \y2=der1(y1,x,x) \y3=der2(y1,x,x) y(x)= WIND ZOOM TRACE GRAPH evalF nDer der1 der2 fnInt ▸
Draw the graph of the first derivative (y2) and the second derivative (y3) of *f* using an appropriate window. You can use ZFIT to set the vertical view or experiment until you find a suitable view. The graph to the right uses −4 ≤ *y* ≤ 4.	ROOT X=4.2730999685 y=0

Find the *x*-intercept of the second derivative graph as indicated in this section or find the input of the high point on the first derivative graph (see Section 5.2.1c) to locate the input of the inflection point. (When more than one graph is on the screen, use ▼ to jump between graphs.)

5.3.2 USING THE CALCULATOR TO FIND INFLECTION POINTS
Remember that an inflection point on the graph of a function is a point of greatest or least slope. Whenever finding the second derivative of a function is tedious algebraically and/or you do not need an exact answer from an algebraic solution, you can easily find the input location of an inflection point by finding where the first derivative of the function has a maximum or minimum slope.

We illustrate this method using the logistic function for polio cases that is in Example 2 of Section 5.3 in *Calculus Concepts*:

The number of polio cases in the U.S. in 1949 is given by $C(t) = \dfrac{42{,}183.911}{1 + 21{,}484.253e^{-1.248911t}}$

where *t* = 1 in January, *t* = 2 in February, and so forth.

Enter $C(x) = 42183.911/(1 + 21484.253e^{\wedge}(-1.248911x))$ in the y1 location of the y(x)= list and the first derivative of *f* in y2. (You can use your algebraic formula for the first derivative or the calculator's derivative.) Turn off y1.	Plot1 Plot2 Plot3 \y1=42183.911/(1+214... \y2=der1(y1,x,x) y(x)= WIND ZOOM TRACE GRAPH x y INSf DELf SELCT▸
The problem context says that the input interval is from 0 (the beginning of 1949) to 12 (the end of 1949), so set these values for *x* in the window. Set the vertical view and draw the graph of *C′* with ZFIT.	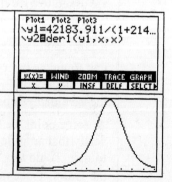

| Use the method discussed in Section 5.2.1c, *i.e.*, MORE F1 [MATH] F5 [FMAX], to find the input location of the maximum point on the slope graph. | 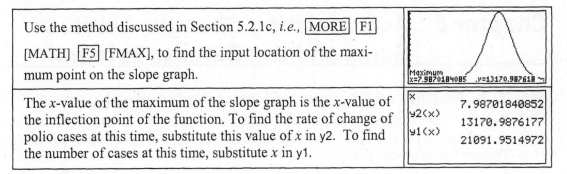 |
| The *x*-value of the maximum of the slope graph is the *x*-value of the inflection point of the function. To find the rate of change of polio cases at this time, substitute this value of *x* in y2. To find the number of cases at this time, substitute *x* in y1. | |

CAUTION: Do not forget to round your answers appropriately (this function should be interpreted discretely) and to give units of measure with each answer.

| Note that you could have found the input of the inflection point on the polio cases graph by finding the *x*-intercept of the second derivative graph. The function whose graph is shown to the right is y3 = der2(y1, x, x), and the graph was drawn using ZFIT. | 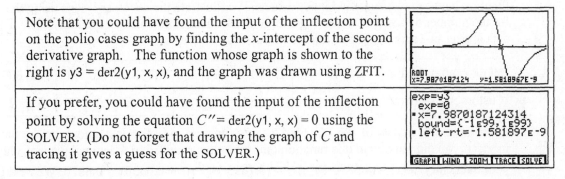 |
| If you prefer, you could have found the input of the inflection point by solving the equation $C'' = $ der2(y1, x, x) = 0 using the SOLVER. (Do not forget that drawing the graph of C and tracing it gives a guess for the SOLVER.) | |

Our final method is also the simplest – using the TI-86's built-in inflection point finder! Be certain if you use this method that the function does have an inflection point at the location indicated by the calculator. We illustrate with the polio cases function that is entered in y1:
$$C(x) = 42183.911/(1 + 21484.253e^{(-1.248911x)}).$$

Turn y1 on and draw a graph of C between $x = 0$ and $x = 12$ with ZFIT. Press MORE F1 [MATH] MORE F1 [INFLC]. Move, using ▶ or ◀, the cursor to a position to the left of where the function changes concavity.	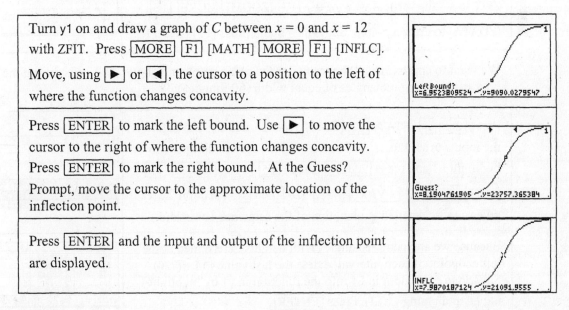
Press ENTER to mark the left bound. Use ▶ to move the cursor to the right of where the function changes concavity. Press ENTER to mark the right bound. At the Guess? Prompt, move the cursor to the approximate location of the inflection point.	
Press ENTER and the input and output of the inflection point are displayed.	

Chapter 6   Accumulating Change: Limits of Sums and the Definite Integral

 6.1 Results of Change and Area Approximations

So far, we have used the TI-86 to investigate rates of change. In this chapter we consider the second main topic in calculus – the accumulation of change. You calculator has many useful features that will assist in your investigations of the results of change.

6.1.1 AREA APPROXIMATIONS USING LEFT RECTANGLES   The TI-86 lists can be used to approximate, using left rectangles, the area between the horizontal axis and a rate of change function between two specified input values. We illustrate the necessary steps using the data for the number of customers entering a large department store during a Saturday sale. These data appear in Table 6.1 of Section 6.1 in *Calculus Concepts*.

Minutes after 9 a.m.	0	45	75	120	165	195	255	330	370	420	495	570	630	675
Customers per minute	1	2	3	4	4	5	5	5	5	4	4	3	2	2

Enter these data in lists L1 and L2. Find a cubic function to fit the data, and paste the function in y1. If you did not obtain this function (shown below rounded to 4 decimal places), check your data entry: $c(x) = (4.5890 \cdot 10^{-8})x^3 - (7.7813 \cdot 10^{-5})x^2 + 0.0330x + 0.8876$ We next graph the function on a scatter plot of the data over the 12-hour (720 minute) sale.

Delete any functions in the y(x)= list and turn on Plot 1. Press [F2] [WIND], either set xMin = 0, xMax = 720, xScl = 60 and use ZFIT to set the vertical view or use [F3] [ZOOM] [MORE] [F5] [ZDATA] to draw a graph. Set yMin = 0 and press [F5] [GRAPH].

We want to approximate the area under the cubic graph and above the *x*-axis during the 12-hour period using rectangles of equal width (60 minutes).

Press [2nd] [+] (STAT) [F2] [EDIT], clear L1 and L2, and enter the inputs 0, 60, 120, ..., 720 in L1. A quick way to do this is to highlight L1 and press [2nd] [–] (LIST) [F5] [OPS] [MORE] [F3] [seq(] [x-VAR] [,] [x-VAR] [,] 0 [,] 720 [,] 60 [)]. (Section 2.1.1a gives a more detailed explanation of the sequence command.)

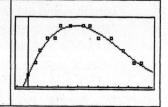

Because we are using rectangles with heights determined at the left endpoint of each interval, delete the last value in L1 (720). Substitute each input in L1 into the function in y1 by highlighting L2 and typing y1(L1). Press [ENTER].

Consider what is now in the lists. L1 contains the left endpoints of the 12 rectangles and L2 contains the heights of the rectangles. If we multiply the heights by the widths of the rectangles (60 minutes) and enter this product in L3, we will have the rectangle areas in L3.

Highlight L3 and type 60 $\boxed{\times}$ $\boxed{\text{ALPHA}}$ 7 (L) 2 $\boxed{\text{ENTER}}$. L3 now contains the areas of the 12 rectangles.

L1	L2	L3	3
0	.8876301	--------	
60	2.599374		
120	3.81034		
180	4.580002		
240	4.967836		
300	5.033313		

L3 =60*L2∎

L1	L2	L3	3
0	.8876301	53.25781	
60	2.599374	155.9624	
120	3.81034	228.6204	
180	4.580002	274.8001	
240	4.967836	298.0701	
300	5.033313	301.9988	

L3(1) =53.257805265234

All that remains is to add the areas of the 12 rectangles. To do this, return to the home screen and type $\boxed{\text{2nd}}$ $\boxed{-}$ (LIST) $\boxed{\text{F5}}$ [OPS] $\boxed{\text{MORE}}$ $\boxed{\text{F1}}$ [sum] $\boxed{\text{ALPHA}}$ 7 (L) 3 $\boxed{\text{ENTER}}$.

We estimate that 2574 customers came to the Saturday sale.

```
sum L3
              2573.71231854
```

6.1.2 **AREA APPROXIMATIONS USING RIGHT RECTANGLES**   When you use left rectangles to approximate the results of change, the rightmost data point is not the height of a rectangle and is not used in the computation of the left-rectangle area. Similarly, when using right rectangles to approximate the results of change, the leftmost data point is not the height of a rectangle and is not used in the computation of the right-rectangle area. We illustrate the right-rectangle approximation using the function r that is given in Example 2 of Section 6.1.

The rate of change of the concentration of a drug in the bloodstream is modeled by

$$r(t) = \begin{cases} 1.708(0.845^x) & \text{when } 0 \le x \le 20 \\ -10.058 + 2.94 \ln x & \text{when } 20 < x \le 30 \end{cases}$$

where x is the number of days after the drug is first administered.

NOTE: It is not necessary to draw a graph of r, but refer to Section 1.2.3 of this *Guide* if you want to review the instructions for graphing this piecewise continuous function. Place the TI-86 in DrawDot mode (also discussed in Section 1.2.3) if you intend to draw a graph of r.

Enter the 2 pieces of r in the y(x)= list as shown on the right. With the cursor in y3, enter $\boxed{(}$ $\boxed{\text{2nd}}$ $\boxed{\text{ALPHA}}$ (alpha) 0 (Y) 1 $\boxed{)}$ $\boxed{(}$ $\boxed{\text{x-VAR}}$ $\boxed{\text{2nd}}$ 2 (TEST) $\boxed{\text{F4}}$ [≤] 20 $\boxed{)}$ $\boxed{+}$ $\boxed{(}$ $\boxed{\text{2nd}}$ $\boxed{\text{ALPHA}}$ (alpha) 0 (Y) 2 $\boxed{)}$ $\boxed{(}$ $\boxed{\text{x-VAR}}$ $\boxed{\text{2nd}}$ 2 (TEST) $\boxed{\text{F3}}$ [>] 20 $\boxed{)}$.

Turn off y1 and y2. (See Section 1.2.2b of this *Guide* for instructions on turning functions in the Y= list off and on.)

```
Plot1 Plot2 Plot3
\y1=1.708(.845^x)
\y2=-10.058+2.94ln x
\y3■(y1)(x≤20)+(y2)(...
```

Part *a* of Example 2 says to find the change in the drug concentration from $x = 0$ through $x = 20$ using right rectangles of width 2 days. Enter the right endpoints (2, 4, 6, ..., 20) in L1.

Use the piecewise continuous function r in y3 (or use y1) to find the rectangle heights. As shown, enter the heights in L2.

L1	L2	L3	2
2	1.219555	--------	
4	.8707925		
6	.6217676		
8	.4439576		
10	.3169969		
12	.2263437		

L2 =y3(L1)∎

- Note that the heights in L2 are the values in Table 6.3 in the text, but the Table 6.3 values have been rounded for printing purposes.

Find the right-rectangle areas by multiplying each entry in L2 by 2. This is the same as multiplying the sum of the heights by 2. On the home screen, press 2 $\boxed{\times}$ $\boxed{\text{2nd}}$ $\boxed{-}$ (LIST) $\boxed{\text{F5}}$ [OPS] $\boxed{\text{MORE}}$ $\boxed{\text{F1}}$ [sum] $\boxed{\text{ALPHA}}$ 7 (L) 2 $\boxed{\text{ENTER}}$.

```
2*sum L2
              8.23530983841
```

Part *b* of Example 2 asks us to use the model and right rectangles of width 2 days to estimate the change in drug concentration from $x = 20$ to $x = 30$. Notice that the signed heights in L4 are the same as those given in the text in Table 6.4 (except for rounding.)

Enter the right endpoints (22, 24, 26, 28, 30) in L3. Use the piecewise continuous function *r* in y3 (or use y2) to find the rectangle heights. As shown to the right, enter the signed heights in L4.	**L2 / L3 / L4** 1.219555 / 22 / -.970335 .8707925 / 24 / -.714522 .6217676 / 26 / -.479196 .4439576 / 28 / -.261319 .3169969 / 30 / -.05848 .2263437 / ------ / ------ L4 =y3(L3)
Find the signed right-rectangle areas by multiplying each entry in L2 by 2. This is the same as multiplying the sum of the signed heights by 2. On the home screen, press 2nd ENTER (ENTRY) to return the sum instruction to the screen, change L2 to L4 with ◄ and then type 4. Press ENTER.	2*sum L2 　　　8.23530983841 2*sum L4 　　　-4.96770308424

6.1.3a AREA APPROXIMATIONS USING MIDPOINT RECTANGLES

Areas of midpoint rectangles are found using the same procedures as those used to find left and right rectangle areas except that the midpoint of the base of each rectangle is entered in the first list and no endpoints are deleted. We illustrate the midpoint-rectangle approximation using the function *f* in Example 3 of Section 6.1 in *Calculus Concepts*.

Clear all functions from the y(x) = list. Next, enter the function $f(x) = \sqrt{1 - x^2}$ in some location of the y(x) = list, say y1. (Type the square root symbol with 2nd x^2.)	Plot1 Plot2 Plot3 \y1圓√(1-x²)
Clear L1 and L2. Using 4 midpoint rectangles to approximate the area of the region between the graph of *f* and the *x*-axis between $x = 0$ and $x = 1$, we first enter the midpoints of the rectangles (0.125, 0.375, 0.625, 0.875) in L1. As shown to the right, enter the heights of the rectangles in L2.	**L1 / L2 / L3** .125 / .9921567 / ------ .375 / .9270248 / ------ .625 / .7806247 .875 / .4841229 L2 =y1(L1)
You can now multiply each rectangle height by the width 0.25 (entering these values in another list) and sum the rectangle areas or you can multiply the sum of the heights in L2 by 0.25. Either calculation results in the area estimate.	.25*sum L2 　　　.795982305153

6.1.3b SIMPLIFYING AREA APPROXIMATIONS

The procedures used in the previous three sections of this *Guide* can become tedious when the number of rectangles is large. When you have a function $y = f(x)$ in the y1 location of the y(x)= list, you will find program NUMINTGL very helpful in determining left-rectangle, right-rectangle, and midpoint-rectangle approximations for accumulated change. Program NUMINTGL (listed in the *TI-86 Program Appendix*) performs automatically all the calculations that you have been doing manually using the lists.

WARNING: This program will not execute properly unless the function that determines the heights uses x as the input variable and is pasted in y1. If you receive an error message while attempting to run this program, consult Programs in *Troubleshooting the TI-86* in this *Guide*.

We illustrate using this program with the function $f(x) = \sqrt{1 - x^2}$ that is used in Example 3 in Section 6.1. This function should be entered in the y1 location of the y(x) = list.

Press ⌐PRGM⌐ ⌐F1⌐ [NAMES] and the F-key under NUMIN. Press ⌐ENTER⌐ to start program NUMINTGL. If you did not enter the function in y1, press ⌐F2⌐ [No] to exit the program. Enter the function in y1 and re-run the program. If the function is in y1, press ⌐F1⌐ [Yes] to continue.	`Enter f(x) in y1`
	`Continue?`
	
	`Yes | No`

At the next prompt, pressing ⌐F1⌐ [Yes] draws the approximating rectangles and pressing ⌐F2⌐ [No] obtains only numerical approximations to the area between the function and the horizontal axis between 0 and 1. For illustration purposes, let's choose to see the rectangles, so press ⌐F1⌐.	`Enter f(x) in y1`
	`Continue?`
	`Draw Pictures?`
	
	`Yes | No`

At the Left endpoint? prompt, press 0 ⌐ENTER⌐, and at the right endpoint? prompt, press 1 ⌐ENTER⌐ to tell the TI-86 the input interval. You are next shown a menu of choices. Press 4 and ⌐ENTER⌐ for a midpoint-rectangle approximation.	`Enter Choice:`
	`Left Rect   (1)`
	`Right Rect  (2)`
	`Trapezoids  (3)`
	`Midpt Rect  (4) 4■`

Input 4 at the N? prompt and press ⌐ENTER⌐. A graph of the 4 midpoint rectangles and the function are shown. (Note that the program automatically sets the height of the window based on the left and right endpoints of the input interval.)	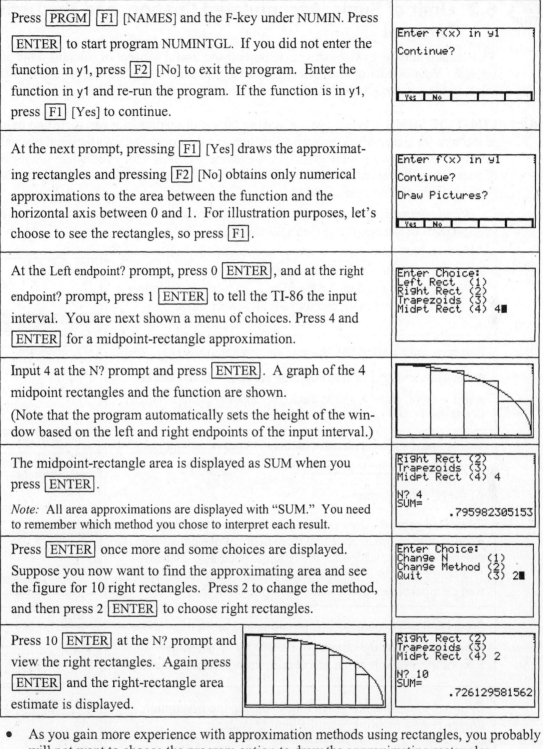

The midpoint-rectangle area is displayed as SUM when you press ⌐ENTER⌐.
 	
Note: All area approximations are displayed with "SUM." You need to remember which method you chose to interpret each result.
	`Right Rect  (2)`
	`Trapezoids  (3)`
	`Midpt Rect  (4) 4`
	
	`N? 4`
	`SUM=`
	`        .795982305153`

Press ⌐ENTER⌐ once more and some choices are displayed.
 	
Suppose you now want to find the approximating area and see the figure for 10 right rectangles. Press 2 to change the method, and then press 2 ⌐ENTER⌐ to choose right rectangles.
	`Enter Choice:`
	`Change N       (1)`
	`Change Method (2)`
	`Quit          (3) 2■`

Press 10 ⌐ENTER⌐ at the N? prompt and view the right rectangles. Again press ⌐ENTER⌐ and the right-rectangle area estimate is displayed.	`Right Rect  (2)`
	`Trapezoids  (3)`
	`Midpt Rect  (4) 2`
	
	`N? 10`
	`SUM=`
	`        .726129581562`

- As you gain more experience with approximation methods using rectangles, you probably will not want to choose the program option to draw the approximating rectangles.

Continue in this manner to find the left-rectangle area or change N and find the left-, right-, or midpoint-rectangle approximations for different numbers of subintervals. When you finish, press ⌐ENTER⌐ and choose 3 to exit the program.	`Enter Choice:`
	`Change N       (1)`
	`Change Method (2)`
	`Quit          (3) 3■`

6.2 Limit of Sums, Accumulated Change, and the Definite Integral

This section introduces you to a very important and useful concept of calculus – the definite integral. Your calculator can be very helpful as you study definite integrals and how they relate to the accumulation of change.

6.2.1 LIMIT OF SUMS When you are looking for a limit in midpoint-rectangle approximations of the area (or signed area) between a function and the horizontal axis between two values of the input variable, program NUMINTGL is extremely useful! However, when finding a limit of sums using the values displayed by this program, it is not advisable to draw pictures when N, the number of rectangles, is large.

We illustrate using this program to find a limit of sums using $f(x) = \sqrt{1-x^2}$, the function in Example 1 of Section 6.2 of *Calculus Concepts*.

To construct a table of midpoint-rectangle approximations for the area between f and the x-axis from $x = 0$ to $x = 1$, first clear the y(x)= list, enter f in y1, and turn off all scatter plots.	Plot1 Plot2 Plot3 \y1◼√(1-x²) y(x)= WIND ZOOM TRACE GRAPH x | y | INSf | DELf | SELCTM
Run program NUMINTGL. At the first prompt, press ☐F1 [Yes] to continue and ☐F2 [No] to not draw pictures. At the Left endpoint? prompt, enter 0, and at the Right endpoint? prompt, enter 1 to tell the TI-86 that the input interval is from $x = 0$ to $x = 1$.	Left endpoint? 0 Right endpoint? 1◼
You are next shown a menu of choices. <u>Always use midpoint rectangles with a limit of sums.</u> Press 4 and ENTER. At the N? prompt, enter 4.	Enter Choice: Left Rect (1) Right Rect (2) Trapezoids (3) Midpt Rect (4) 4 N? 4◼
Record on paper the area estimate for 4 midpoint rectangles. (You should record at least one more decimal place than the number of places needed for the required accuracy. Because Example 1 asks for 3-decimal-place accuracy, record at least four decimal places of the value shown to the right.)	Right Rect (2) Trapezoids (3) Midpt Rect (4) 4 N? 4 SUM= .795982305153
Press ENTER to continue the program, and press 1 to choose Change N. Double the number of rectangles by entering N = 8. Record the area approximation (again, to four decimal places).	Change N (1) Change Method (2) Quit (3) 1 N? 8 SUM= .789171732825
Continue in this manner, each time choosing the first option, Change N, and doubling N until a limit is evident. When you finish, press ENTER and choose 3 to exit the program.	Change N (1) Change Method (2) Quit (3) 1 N? 256 SUM= .78541918062

NOTE: Intuitively, *finding the limit* means that you are sure what the area approximation, to 3 decimal places (or whatever accuracy is specified in the problem), will be without having to use larger values of N in the program.

6.4 The Fundamental Theorem

The Fundamental Theorem of Calculus is important because it connects the two main topics in calculus – differentiation and integration. In this section of the *Guide*, we see how to use the TI-86's definite integral function and use it along with the TI-86's numerical derivative to illustrate the Fundamental Theorem in action.

6.4.1a DEFINITE INTEGRAL NOTATION AND CALCULATOR NOTATION Recall that der1 and nDer are the TI-86's derivatives and provide, in most cases, the value of or a good approximation for the instantaneous rate of change of a function when that rate of change exists. As is the case with the TI-86 derivatives, your calculator does not give a formula for accumulation functions. However, it does give an excellent numerical estimate for the definite integral of a function between two specific input values when that integral exists.

The TI's numerical integral is called fnInt, and the correspondence between the mathematical notation $\int_a^b f(x)\,dx$ and the calculator's notation fnInt(*f*(x), x, a, b) is illustrated in Figure 6.

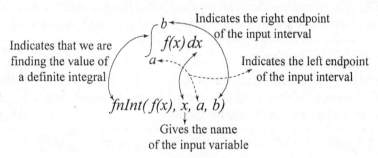

Figure 6

We illustrate the use of the calculator's definite integral function by finding the area from $x = 0$ to $x = 1$ between the function $f(x) = \sqrt{1-x^2}$ and the horizontal axis. Recall that the limit (to 3 decimal places) found from the limit of sums investigation in Example 1 of Section 6.2 of *Calculus Concepts* was 0.785. (Also see Section 6.2.1 of this *Guide*.)

You can enter *f* in y1 and refer to it in the fnInt expression as y1. Access fnInt with [2nd] [÷] (CALC) [F5] [fnInt].	Plot1 Plot2 Plot3 \y1◼√(1-x²) y(x)= WIND ZOOM TRACE GRAPH x y INSf DELf SELCT▶	fnInt(y1,x,0,1) .785398169531 evalF nDer der1 der2 fnInt

NOTE: Any y(x)= list location can be used to hold the function. Recall that when in the y(x)= list, you must use x as the input variable name.

Or, you can enter the function directly on the home screen as shown to the right. When you type a function formula on the home screen, you can use any letter as the input variable symbol.	fnInt(√(1-x²),x,0,1) .785398169531 fnInt(√(1-T²),T,0,1) .785398169531

• The TI-83's fnInt function yields the same result (to 3 decimal places) as that found in the limit of sums investigation in Section 6.2.1 of this *Guide*.

6.4.1b THE FUNDAMENTAL THEOREM OF CALCULUS Intuitively, this theorem tells us that the derivative of an antiderivative of a function is the function itself. Let us view this idea both numerically and graphically. The correct syntax for the TI-86's numerical integrator is fnInt(function, name of input variable, left endpoint for input, right endpoint for input)

Consider the function $f(t) = 3t^2 + 2t - 5$ and the accumulation function $F(x) = \int_1^x f(t)\,dt$.

The Fundamental Theorem of Calculus tells us that $F'(x) = \dfrac{d}{dx}\left(\int_1^x f(t)\,dt\right) = f(x)$; that is, $F'(x)$ is f evaluated at x.

Input f in y1 and $F' = $ nDer(fnInt(y1, x, 1, x), x, x) in y2 (remember that the TI-86 requires that you use x as the input variable in the y(x)= list). Access nDer with [2nd] [÷] (CALC) [F2] [nDer] and fnInt with [2nd] [÷] (CALC) [F5] [fnInt]. (Turn off any stat plots that are turned on.)	Plot1 Plot2 Plot3 \y1■3x²+2x-5 \y2■nDer(fnInt(y1,x,… x y INSf DELf SELCT evalF nDer der1 der2 fnInt▶

NOTE: The numerical derivative (nDer) is used rather than the function derivative (der1) because fnInt is not one of the pre-programmed functions for which der1 can be calculated.

Press [TABLE] [F2] [TBLST], set Indpnt: to ASK, and press [F1] [TABLE]. Input several different values for x. Other than some occasional round-off error because the calculator is approximating these values, the results are identical.	x y1 y2 -5 60 60 4.3 59.07 59.07 159.82 76941.94 76941.94 ■1.6667 6.667067 6.667068 x=1.6667 TBLST SELCT x y
Find a suitable viewing window such as when you use [GRAPH] [F3] [ZOOM] [F4] [ZSTD]. Without changing the window (that is, draw the graphs by pressing [F5] [GRAPH]), turn off y2 and draw the graph of y1. Then turn off y1 and draw the graph of y2.	y(x)= WIND ZOOM TRACE GRAPH▶

Note that the graph of y2 takes a while to draw. This is because the TI-86 is not only plotting the points, it is also calculating the numerical integral and then the numerical derivative before plotting each point. Turn both y1 and y2 on and draw the graph of both functions. Only one graph is seen in each case.

EXPLORE: Enter several other functions in y1 and do not change y2 except possibly for the left end-point 1 in the fnInt expression. Perform the same explorations as above. Confirm your results with derivative and integral formulas.

6.4.2 DRAWING ANTIDERIVATIVE GRAPHS   Recall when using fnInt($f(x)$, x, a, b) that a and b are, respectively, the lower and upper endpoints of the input interval. Also remember that you do not have to use x as the input variable unless you are graphing the integral or evaluating it using the calculator's table. Unlike when graphing using der1 or nDer, the TI-86 will not graph a general antiderivative; it only draws the graph of a specific accumulation function. Thus, we can use x for the input at the upper endpoint when we want to draw an antiderivative graph, but not for the inputs at both the upper and lower endpoints.

All of the antiderivatives of a specific function differ only by a constant. We explore this idea using the function $f(x) = 3x^2 - 1$ and its general antiderivative $F(x) = x^3 - x + C$. Because we are working with a general antiderivative in this illustration, we do not have a starting point for the accumulation. We therefore choose some value, say 0, to use as the starting point for the accumulation function to illustrate drawing antiderivative graphs. If you choose a different lower limit, your results will differ from those shown below by a constant.

Enter f in y1, fnInt(y1, x, 0, x) in y2, and F in y3, y4, y5, and y6, using a different number for C in each function location. (You can use the values of C shown to the right or different values.)	Ploti Plot2 Plot3 \y1□3x²-1 \y2□fnInt(y1,x,0,x) \y3□x^3-x+1 \y4□x^3-x-2 \y5□x^3-x+5 \y6□x^3-x-4.8 y(x)= WIND ZOOM TRACE GRAPH
Find a suitable viewing window and graph the functions y1 through y6. The graph to the right was drawn with $-3 \le x \le 3$ and $-20 \le y \le 20$.	
It appears that the only difference in the graphs of y2 through y6 is the y-axis intercept. But, isn't C the y-axis intercept of each of these antiderivative graphs? Delete y4, y5, and y6. Turn off y1 and change the 1 in y3 to 0.	Ploti Plot2 Plot3 \y1=3x²-1 \y2□fnInt(y1,x,0,x) \y3□x^3-x+0 y(x)= WIND ZOOM TRACE GRAPH x y INSF DELF SELCT
Press EXIT F5 [GRAPH] and draw the graphs of y2 and y3. You should see only one graph. Set the TI-86 TABLE to ASK and enter some values for x. It appears that y2 and y3 are the same function.	x y2 y3 -5 -120 -120 0 0 0 -4.3 75.207 75.207 -8.35 -573.833 -573.833 21 9240 9240 51.2 134166.5 134166.5 x=51.2 TBLST SELCT x y

CAUTION: The methods for checking derivative formulas that were discussed in Sections 4.3.2b and 4.3.2c are not valid for checking general antiderivative formulas. Why not? Because to graph an antiderivative using fnInt, you must arbitrarily choose values for the constant of integration and for the input of the lower endpoint. However, for most of the rate-of-change functions where $f(0) = 0$, the calculator's numerical integrator values and your antiderivative formula values should differ by the same constant at every input value where they are defined.

6.5 The Definite Integral

When using the numerical integrator on the home screen, enter fnInt($f(x)$, x, a, b) for a specific function f with input x and specific values of a and b. (Remember that the input variable does not have to be x when the function formula is entered on the home screen.) If you prefer, f can be in the y(x)= list and referred to as y1 (or whatever location is chosen) when using fnInt.

6.5.1a EVALUATING A DEFINITE INTEGRAL ON THE HOME SCREEN

We illustrate the use of fnInt with the function that models the rate of change of the average sea level. The rate-of-change data are given in Table 6.18 of Example 3 in Section 6.5 of *Calculus Concepts*.

Time (thousands of years before the present)	−7	−6	−5	−4	−3	−2	−1
Rate of change of average sea level (meters/year)	3.8	2.6	1.0	0.1	−0.6	−0.9	−1.0

Enter the time values in list L1 and the rate of change of the average sea level values in list L2.

A scatter plot of the data indicates a quadratic function. Find the function and paste it in y1. Draw the function on the scatter plot of the data to confirm that it gives a good fit.	QuadraticReg y=ax²+bx+c n=7 PRegC= {.147619047619 .3595... CALC EDIT PLOT DRAW VARS PwrR SinR L9stR P2Re9 P3Re9
Because part a of Example 3 asks for the *areas* of the regions above and below the input axis and the function, we must find where the function crosses the axis. You can find this value using the solver (solve y1 = 0) or by using the graph and the x-intercept method described in Section 1.1.1i of this *Guide*.	ROOT x=-3.844955338 y=-1E-13

CAUTION: If you round the *x*-intercept value, it will cause whatever you do with this value to not give as an accurate result as possible. This situation occurs many times in this and the next several chapters of the text. Also recall that one of the numerical considerations given in Chapter 1 of *Calculus Concepts* is that intermediate calculation values are not to be rounded. For maximum accuracy, store this value in some memory location, say Z, and refer to it as Z in all subsequent calculations.

Whether you use the SOLVER or the *x*-intercept method shown in the last calculation, the zero is stored in x. Return to the home screen and store the value in Z (or whatever memory location, except x, that you choose). Do not use the x location because the x value changes whenever you use the TI-86 table, the solver, or trace a graph.

Find the area of the region above the horizontal axis by typing and entering the first expression shown on the right.
 	
 Find the area of the region below the horizontal axis by typing and entering the second expression shown on the right. (The negative is used because the region is below the input axis.)
	`x→Z`
	 `              -3.84495533758`
	 `fnInt(y1,x,-7,Z)`
	 `              5.40593487597`
	 `-fnInt(y1,x,Z,0)`
	 `              2.93649043153`
	
	 `evalF  nDer  der1  der2  fnInt ▶`

Part *b* of Example 3 asks you to evaluate $\int_{-7}^{0} y1\, dx$. Find this value by typing and entering the expression shown on the right.
 	
 Note that the result is not the sum of the two areas – it is their difference.
	`fnInt(y1,x,-7,0)`
	 `              2.46944444444`
	
	 `evalF  nDer  der1  der2  fnInt ▶`

- If you evaluate a definite integral using antiderivative formulas and check your answer with the calculator using fnInt, you might find a slight difference in the last few decimal places. Remember that the TI-86 is evaluating the definite integral using an approximation technique, not an algebraic formula.

6.5.1b EVALUATING A DEFINITE INTEGRAL FROM THE GRAPHICS SCREEN The value of a definite integral can also be found from the graphics screen. We again illustrate the use of fnInt with the function that models the rate of change of the average sea level. You should practice using the following method by finding the areas of the regions indicated in part *c* of Example 4 in Section 6.5 of the text.

Turn off Plot 1 and have the function modeling the average sea level in y1. (See Section 6.5.1a of this *Guide*.) We want to evaluate $\int_{-7}^{0} y1\, dx$.	`Plot1 Plot2 Plot3`
	 `\y1⊟.14761904761904x...`
	
	 `y(x)= WIND  ZOOM  TRACE GRAPH`
	 ` x    y   INSf DELf SELCT▶`

Press EXIT F2 [WIND], set xMin = ⁻7, xMax = 0, and draw the graph of y1 with F3 [ZOOM] MORE F1 [ZFIT].	
Press MORE F1 [MATH] F3 [∫f(x)]. The TI-86 asks Lower Limit? Press (−) 7 and obtain the screen shown to the right.	`Lower Limit?` `x=-7▪`
Press ENTER. The TI-86 now asks Upper Limit? Press 0 and obtain the screen shown to the right.	`Upper Limit?` `x=0▪`

Press ENTER . The region between the function and the horizontal axis from the lower limit to the upper limit is shaded. The value of the integral is shown at the bottom of the screen.

CAUTION: If you type in a value for the lower and/or upper limit (that is, the input at the endpoint) that is not visible on the graphics screen, you will get an error message when you attempt to evaluate the integral. Be certain that these endpoints are included in the interval from xMin to xMax before using this method.

6.5.2 FINDING THE AREA BETWEEN TWO CURVES   The process of finding the area of the region between two functions uses many of the techniques presented in preceding sections. If the two functions intersect, you need to first find the input values of the point(s) of intersection. We illustrate these ideas as they are presented in Example 5 of Section 6.5 of the text.

Clear all functions from the y(x)= list and turn off the stat plots. Enter the function $s(t) = 3.7(1.19376^t)$ in y1 and the function $a(t) = 0.04t^3 - 0.54t^2 + 2.5t + 4.47$ in y2. (Remember to use x as the input variable in the y(x)= list.).	
From Figure 6.59 in the text, we see that the regions under discussion use xMin = 0, xMax = 21, yMin = 0, and yMax ≈ 160. Press EXIT F2 [WIND], set these values, and then press F5 [GRAPH]. (Recall that CLEAR removes the menu from the screen.)	
To better view the points of intersection, reset xMax ≈ 16, yMin ≈ ⁻10 and yMax ≈ 60 and press GRAPH . (Recall that EXIT or GRAPH causes the menu to reappear at the bottom of the screen.)	

NOTE: TRACE can be used to estimate these maximum values, and we set yMin to a value less than 0 so that the displayed coordinates do not cover a portion of the graph when tracing.

We next find the inputs of the points of intersection of the two functions. (These values will probably be the limits on the integrals we use to find the areas.) The method we use to find the first of the points is the intersection method that was discussed in Section 1.1.1i of this *Guide*.

With the graph on the screen, press MORE F1 [MATH] MORE F3 [ISECT]. The TI-86 asks First curve? Press ENTER to mark y1. The calculator then asks Second curve? Press ENTER to mark y2. At the Guess? prompt, use ◄ to move the cursor near the leftmost visible point of intersection.	
Press ENTER to find $t ≈ 3.724$. To avoid making a mistake copying this value and to eliminate as much rounding error as possible, return to the home screen and store the input value in A with the keystrokes x-VAR STO▶ LOG (A) ENTER .	

You could use this same method to find B, the input value of the rightmost point of intersection. However, to illustrate another way this can be done, we find this value using the SOLVER.

The equation we are solving is y1 = y2, but remember that what you enter in the SOLVER must have a constant on one side of the equation. Access the SOLVER with 2nd GRAPH (SOLVER). Press CLEAR. Enter y1 – y2 in the eqn location. Press ENTER.	``eqn:y1-y2``
	
	``ReSEq  y1  y2``

Enter 0 for the right-hand side of the equation (in the second exp position). Note from the graph that B is near 14. Use this or some other value closer to B than it is to A as your guess. Press F5 [SOLVE] to find that $B \approx 14.242$.	``exp=y1-y2``
	``  exp=0``
	``•x=14.241965139302``
	``  bound={-1E99,1E99}``
	``•left-rt=-2E-12``
	
	``GRAPH WIND ZOOM TRACE SOLVE``

As with the first point, to avoid making a mistake copying this input value and to eliminate as much rounding error as possible, return to the home screen and store this value in B with the keystrokes x-VAR STO▶ SIN (B) ENTER.	``x→A``
	``            4.65695969986``
	``x→B``
	``           14.2419651393``

NOTE: Even though we have found the only intersection points that are useful in the context of this example, the SOLVER can easily find if there are any other intersection points for these two curves. Enter several different guesses, some smaller than A and some larger than B. (We can see from the graph that there are no other points of intersection between A and B.) You should find that these curves also intersect at $t \approx -0.372$ and $t \approx 28.077$.

Calculate the areas of the two regions enclosed by s and a (that is, regions R_1 and R_2 that are shown in Figure 6.60 in the text) as indicated to the right.	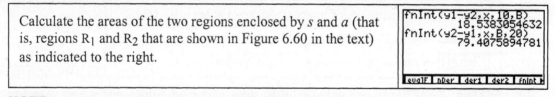

NOTE: Each answer should be positive because we are finding areas. If you obtain a negative answer, you probably entered the functions in the wrong order.

If you are in doubt as to which function to enter first in the integral, press GRAPH F5 [GRAPH] and trace the graphs to see which function is on top of the other. Press ▼ to have the cursor jump from one curve to the other and notice the number of the function's position in the top right corner of the screen.

NOTE: The value $18.53830546 - 79.40758948 \approx -60.869$ could also have been calculated by evaluating fnInt(y1 – y2, x, 10, 20). Because the graph in Figure 6.60 shows that the area of R_2 is greater than the area of R_1, we see that fnInt(y1 – y2, x, 10, 20) should be negative.

6.6 Average Value and Average Rates of Change

Average rates of change are computed as discussed in Section 3.1.1a of this *Guide*. When finding an average value, you need to carefully read the question in order to determine which quantity should be integrated. Considering the units of measure in the context can be a great help when trying to determine which function to integrate to find an average value.

6.6.1a AVERAGE VALUE OF A FUNCTION We illustrate finding an average value with the data in Table 6.21 in Example 1 of Section 6.6 of *Calculus Concepts*:

Time (number of hours after midnight)	7	8	9	10	11	12	13	14	15	16	17	18	19
Temperature (°F)	49	54	58	66	72	76	79	80	80	78	74	69	62

Clear any old data. Delete any functions in the y(x)= list and turn on Plot 1. Enter the data in the above table in lists L1 and L2.

A scatter plot of the data indicates an inflection point (around 9 p.m.) and no limiting values. Fit a cubic function to the data and paste it in y1.		CubicReg y=ax³+bx²+cx+d n=13 PRegC= {-.035256410256 .718...

Part *b* of Example 1 asks for the average *temperature* (*i.e.*, the average value of the temperature) between 9 a.m. and 6 p.m. So, integrate the *temperature* between $x = 9$ and $x = 18$ and divide by the length of the interval to find that the answer is about 74.4°F.		fnInt(y1,x,9,18)
		         669.845061189
		Ans/(18-9)
		         74.427229021
		
		evalF nDer der1 der2 fnInt ▸

6.6.1b GEOMETRIC INTERPRETATION OF AVERAGE VALUE What does the average value of a function mean in terms of the graph of the function? We continue with Example 1 of Section 6.6 of *Calculus Concepts* by considering the function and average value found in Section 6.6.1a. Have the unrounded cubic temperature function in y1.

Enter 74.427229021 in y2. You can either leave yMin at the setting from the scatter plot or change yMin to 0. Press F5 [GRAPH] to see the graph of the average value and the function.		

We illustrate for only the second graph shown above, but what follows is true for both. The area of the rectangle whose height is the average temperature is $(74.427229021 - 0)(18 - 9) \approx$ 669.845. The area of the region between the temperature function and the input axis between 9 a.m. and 6 p.m. is this same value.		74.427229021*(18-9)
		         669.845061189
		fnInt(y1,x,9,18)
		         669.845061189

To find the answer to part *d* of Example 1, enter and evaluate the expression shown to the right. The average rate of change of the temperature from 9 a.m. to 6 p.m. is about 0.98°F/hr.		y1(18)-y1(9)
		         8.85764235764
		Ans/(18-9)
		         .984182484182

Chapter 7 Analyzing Accumulated Change: Integrals in Action

 7.2 Streams in Business and Biology

You will find your calculator very helpful when dealing with streams that are accumulated over finite intervals. Finding either the future or present value of a continuous income stream is simply finding the value of a definite integral. However, the technique used with discrete income streams involves the TI-86's sequence function that was introduced in Section 2.1.1a. The details are presented in Section 7.2.2 of this Guide.

7.2.1 DETERMINING THE FLOW-RATE FUNCTION FOR AN INCOME STREAM The TI-86 can often help you to find the equation for an income stream flow rate. Note that it is *not* necessary to use the calculator to find such an equation – we present this as a technique to use only if you find writing the equation from the word description difficult. We illustrate these ideas as they are given in Example 1 of Section 7.2 of *Calculus Concepts*.

In Example 1, part *a*, we are told that the business's profit remains constant. Clear lists L1 and L2. In L1 enter two possible input values for the time involved. (You might use different years than the ones shown here.) In L2 enter the amount invested: 10% of the constant profit.	L1: 1, 2 L2: 57900, 57900 L3: ------ L2(3) =
You need to remember that a *constant output* means a *linear flow rate*. Fit a linear function to these two data points to find that $R(t) = 57,000$ dollars per year.	LinReg y=a+bx a=57900 b=0 ↓corr=

CAUTION: If you attempt to draw a scatter plot on the calculator, you will get an error message because the TI-86, using the output data in L2, sets yMin = yMax. (You need to draw the scatter plot using paper and pencil.) If you want to see the horizontal line graph on the TI-86, change yMin and yMax so that 57,900 is between the two values and press F5 [GRAPH].

In Example 1, part *b*, we are told that the business's profit grows by $50,000 each year. The first year's profit (which determines the initial investment at $t = 0$) is $579,000. Reason that if the profit grows by $50,000 each year, the next year's profit will be $579,000 + 50,000 = $629,000. Enter these values in L1 and L2.	L1: 0, 1 L2: 579000, 629000 L3: ------ L2(3) =
You need to remember that *constant growth* means a *linear flow rate*. Fit a linear function to these two data points. Next, carefully read the problem once more. Note that only 10% of the profit is invested. Thus, the linear flow rate function is $R(t) = 0.10(50,000t + 579,000)$ dollars per year t years after the first year of business.	LinReg y=a+bx a=579000 b=50000 corr=1 n=2
In Example 1, part *c*, we are told that the business's profit grows by 17% each year. The first year's profit (which determines the initial investment at $t = 0$) is $579,000. Reason that if the profit grows by 17% each year, the next year's profit will be $579,000 + 0.17(579,000) = $677,430. Enter these values in L1 and L2.	L1: 0, 1 L2: 579000, 677430 L3: ------ L2(3) =

You need to remember that a *constant percentage growth* means an *exponential flow rate*. Fit an exponential function to these two data points. Now, carefully read the problem once more. Note that only 10% of the profit is invested. Thus, the exponential flow rate function is $R(t) = 0.10(579,000)(1.17^t)$ dollars per year t years after the first year of business.

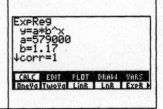

Part *d* of Example 1 gives data that describe the growth of the business's profit. Refer to the material in Section 2.2.1 of this *Guide* to review how to fit a log function to these data points.

NOTE: If you forget which type of growth gives which function, simply use what you are told in the problem and fill in the lists with approximately five data points. Draw a scatter plot of the data and it should be obvious from the shape which function to fit to the data.

For instance, return to Example 1, part *b*, in which we are told that the business's profit grows by $50,000 each year. Note that we assume throughout this section that initial investments are made at time $t = 0$. Simply add $50,000 to each of the previous year's profit to obtain about five data points.

Draw a scatter plot, and observe that the points fall in a line! Note that if you run program DIFF, the first differences are constant at 50,000.

Next, find a linear equation for the data. You should find that the flow rate function is $R(t) = 0.10(50,000t + 579,000)$ dollars per year t years after the first year of business.

7.2.2 FUTURE VALUE OF A DISCRETE INCOME STREAM

The future value of a discrete income stream is found by adding, as d increases, the terms of $R(d)\left(1+\frac{r}{n}\right)^{D-d}$ where $R(d)$ is the value per period of the dth deposit, $100r\%$ is the annual percentage rate at which interest is earned when the interest is compounded once in each deposit period, n is the number of times interest is compounded (and deposits are made) during the year, and D is the total number of deposit periods. It is assumed that initial deposits are made at time $t = 0$ unless it is otherwise stated. We use the situation in Example 4 in Section 7.2 of *Calculus Concepts*:

> When you graduate from college (say, in 3 years), you would like to purchase a car. You have a job and can put $75 into savings each month for this purchase. You choose a money market account that offers an APR of 6.2% compounded quarterly.

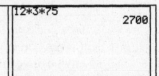

Part *a* asks how much will you have deposited in 3 years. No interest is involved in this calculation. There are 12 months in each year and $75 is deposited each month for 3 years. A total of $2700 is deposited.

Part *b* of Example 4 asks for the future value of the deposits at the end of 3 years. Because the APR is 6.2% compounded quarterly, $r = 0.062$, $n = 4$, and $D = 3$ years $\cdot 4 \frac{\text{deposits}}{\text{year}} = 12$ deposits. A constant $75 each month is deposited, so $R(d) = \$75(3) = \$225$ deposited each quarter. The 3-year future value is given by $\sum_{d=0}^{11} 225\left(1+\frac{0.062}{4}\right)^{12-d}$.

The TI-86 sequence command finds this sum. The syntax for this sequence command is

seq(formula, variable, first value, last value, increment)

When applied to discrete income streams, the *formula* for the sequence is $R(d)\left(1+\dfrac{r}{n}\right)^{D-d}$.

For convenience, we enter the formula in y1. You must enclose the exponent in parentheses. Otherwise, the TI-86 will assume that only 12 is in the exponent.	Plot1 Plot2 Plot3 \y1◻...+.062/4)^(12-x) y(x)= WIND ZOOM TRACE GRAPH x y INSf DELf SELCT▸
Return to the home screen and enter seq(y1, x, 0, 11, 1). Recall that the sequence command is accessed by pressing 2nd − (LIST) F5 [OPS] MORE F3 [seq]. If you want to see the 12 future values in this list, scroll through the list with ▶.	seq(y1,x,0,11,1) {270.608635235 266.4... { } NAMES EDIT OPS sum prod seq liʰʋc vcʰli▸

NOTE: When you use the sequence command for discrete income streams, the first value is always 0. The last value is always $D-1$ because we start counting at 0, not 1. The increment will always be 1 because of the way the formula is designed.

What are the values in the list that result when you use the sequence command? The first value (approximately $270.61) is the 3-year future value of the first 3 months deposits ($225). The second value (approximately $266.48) is the 3-year future value of the second 3 months of deposits ($225), and so forth.

To find the future value of all the deposits, simply sum the values in the list. To do this, use the sum command to add the values in the list (the previous ANS). The sum command is accessed from the LIST OPS screen with F1 [sum]. Store this value in F for use in the next part of the problem.	{270.608635235 266.4...
	sum Ans
	     2988.10123102
	Ans→F
	     2988.10123102
	
	{ } NAMES EDIT OPS
	sum prod seq liʰʋc vcʰli▸

Part *c* of Example 4 asks for the present value of the amount you would have to deposit now to achieve the same future value as was found above. We want to find the present value *P*.

We now solve the equation $P\left(1+\dfrac{r}{n}\right)^{D}$ = Future value. Press 2nd GRAPH (SOLVER), and enter the equation $P\left(1+\dfrac{0.062}{4}\right)^{12}$ = F. Press ENTER .	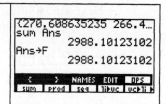exp=P(1+.062/4)^12 exp=F P=2000 bound={-1ᴇ99,1ᴇ99} GRAPH WIND ZOOM TRACE SOLVE
Enter a guess for *P*. Note that F converts to its decimal equivalent when you move to the next line. (There is only one solution so any positive number should work as a guess.) With the cursor on the *P* line, press F5 [SOLVE] to find $P \approx \$2484.48$.	exp=P(1+.062/4)^12 exp=2988.1012310233 •P=2484.4838243874 bound={-1ᴇ99,1ᴇ99} •left-rt=0 GRAPH WIND ZOOM TRACE SOLVE
Part *d* of Example 4 asks for the future value if the monthly interest rate is 0.5% instead of the previous APR compounded quarterly. Press GRAPH F1 [y(x)=] and make the appropriate changes in the function in y1. Don't forget that $R(d)$ is the amount deposited per interest compounding period.	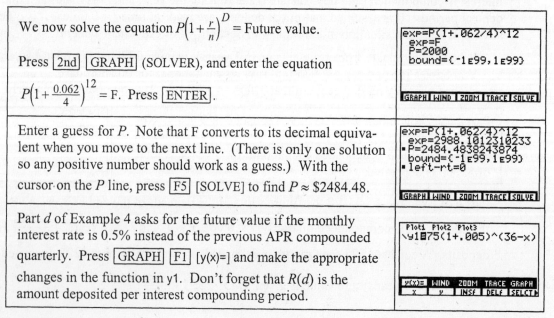Plot1 Plot2 Plot3 \y1◻75(1+.005)^(36-x) y(x)= WIND ZOOM TRACE GRAPH x y INSf DELf SELCT▸

Press [2nd] [ENTER] (ENTRY) until the sequence command re-appears on the home screen. Change *last value* to 35 and press [ENTER]. Sum the resulting sequence with [F1] [sum] [2nd] [(–)] (ANS) to find that the 3-year future value is about $2964.96.

NOTE: You can use any letter for the input variable in the formula if you enter the function on the home screen, but if you are entering the formula in the y(x)= list, you must use x.

7.3 Integrals in Economics

Consumers' surplus and producers' surplus (when they are defined by definite integrals) are easy to find using fnInt. You should draw graphs of the demand and supply functions and think of the economic quantities in terms of area so to understand the questions being asked.

7.3.1 CONSUMER ECONOMICS

We illustrate how to find the consumers' surplus and other economic quantities when the demand function intersects the input axis as given in Example 1 of Section 7.3 of *Calculus Concepts*:

Suppose the demand for a certain model of minivan in the United States can be described as $D(p) = 14.12(0.933^p) - 0.25$ million minivans when the market price is p thousand dollars per minivan.

We first draw a graph of the demand function. This is not asked for, but it will really help your understanding of the problem. Read the problem to see if there are any clues as to how to set the horizontal view for the graph. The price cannot be negative, so $p \geq 0$. There is no price given in the remainder of the problem, so just guess a value with which to begin.

Enter D in y1, using x as the input variable. Press [EXIT] [F2] [WIND], set xMin = 0 and we choose xMax = 20 (remember that the price is in thousands of dollars). Use [F3] [ZOOM] [MORE] [F1] [ZFIT] to graph. Reset yMin = 0 and press [F5] [GRAPH].	
Even though D is an exponential function, a constant has been subtracted from the exponential portion. So, D may cross the input axis. Notice that if it does, the x-intercept will be greater than 20. You could try different values for xMax, but we choose to use the SOLVER to see if there is a value such that y1 = 0.	exp=y1 exp=0 ■x=58.167008627509 bound={-1E99,1E99} ■left-rt=-1E-14 GRAPH WIND ZOOM TRACE SOLVE
Customers will not purchase this model minivan if the price per minivan is more than about $58.2 thousand. Store this value in P for later use and set xMax = P. Redraw the graph with [F5] [GRAPH]. (Be sure to label the axes with variables and units of measure when you copy this graph to your paper.) Note that the answer to Example 1, part c, is $p \approx \$58.2$ thousand.	

Part a of Example 1 asks at what price consumers will purchase 2.5 million minivans. Look at the labels on your graph and note that 2.5 is a value of D. You therefore need to find the price (an input).

Return to the SOLVER and edit the equation so that you are solving y1 = 2.5. You can trace the graph for a guess, but there is only one answer, so any reasonable guess will suffice.	`exp=y1`
	`  exp=2.5`
	`■x=23.590331321715`
	`  bound={-1ᴇ99,1ᴇ99}`
	`■left-rt=0`
	`GRAPH WIND ZOOM TRACE SOLVE`

Part *b* asks for the consumers' expenditure when purchasing 2.5 million minivans. First, store this market price in *M* for future use. (Also label this value *M* on your hand-drawn graph.) The consumers' expenditure is *price * quantity* = area of the rectangle with height = 2.5 million minivans and width ≈ \$23.59 thousand per minivan. The area is about \$59 billion.	`x→M`
	`            23.5903313217`
	`Ans*1000`
	`            23590.3313217`
	`Ans*2.5*10^6`
	`           58975828304.3`
	`58.9758283043*10^9`

The consumer's surplus in part *d* of Example 1 is the area under the demand curve to the right of *M* (*M* ≈ 23.5903) and to the left of *P* (*P* ≈ 58.1670). The surplus is about \$27.4 billion.	`M`
	`            23.5903313217`
	`P`
	`            58.1670086275`
	`fnInt(y1,x,M,P)`
	`            27.4048167344`
	`evalF nDer der1 der2 fnInt▸`

The TI-86 draws the consumers' surplus if you use the methods illustrated in Section 6.5.1b. First, reset yMin to ‑2 to have more room at the bottom of the graph. Reset xMax to 58.2 or an error message comes up with the next instructions. Draw the graph of *D* with F5 [GRAPH].

With the graph on the screen, press MORE F1 [MATH] F3 [∫f(x)]. At the Lower Limit? prompt, press ALPHA 8 (M) ENTER. At the Upper Limit? prompt, press ALPHA , (P) ENTER. You should then see the screen shown to the right.	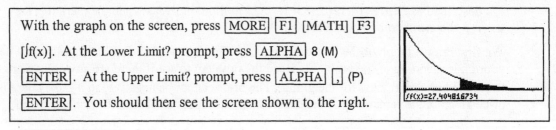

CAUTION: Unless the values you enter for the upper and lower limits are visible on (and not on the boundary of) the graphics screen, you will get an error message using the above instructions. If you do as instructed and use *P*, make sure you used *P* or a value larger than *P* when setting xMax.

7.3.2 **PRODUCER ECONOMICS AND SOCIAL GAIN** We illustrate how to find producers' surplus and other economic quantities as indicated in Example 3 of Section 7.3 in *Calculus Concepts*:

> The demand and supply functions for the gasoline example in the text are given by $D(p) = 5.43(0.607^p)$ million gallons and $S(p) = 0$ million gallons for $p < 1$ and $S(p) = 0.792p^2 - 0.433p + 0.314$ million gallons for $p \geq 1$ when the market price of gas is p dollars per gallon.

First, let's find the market equilibrium point. This point can be found graphically using the intersection command and the graphs of the functions, but we choose to use the SOLVER.

Enter *D* in y1. Enter S (for $p \geq 1$) in y2 as shown to the right. Use x as the input variable in each function. Recall that the piecewise function and its input each need to be in parentheses. Access the inequality symbol with 2nd 2 (TEST) F5 [≥]. Because we intend to draw a graph of *S*, put y2 in DrawDot mode.	

Access the SOLVER and enter the equation y1 − y2 = 0. Enter a guess for the equilibrium point. With the cursor on the x line, press ⌐F5⌐ [SOLVE] to find x ≈ 1.8331. Store this value in some memory location, say *E*.

NOTE: Is x ≈ 1.8331 the only solution to the equation y1 − y2 = 0? Enter more guesses for x, some large and some small. Each time the solution shown above results, so it appears that there is only one solution.

We next draw graphs of the demand and supply functions. How do we set the window to draw the graphs? Notice that because *D* is an exponential function, it will never cross the input axis. Because *S* is a concave-up parabola, it also increases when x > 1. Therefore, just try several different xMax values and choose an xMax so that the graphs can be viewed.

Press ⌐GRAPH⌐ ⌐F2⌐ [WIND], set xMin = 0 and we choose xMax = 3. Use ⌐F3⌐ [ZOOM] ⌐MORE⌐ ⌐F1⌐ [ZFIT] to draw the graphs of *D* and *S*. If yMin is not 0, set it to 0 and press ⌐F5⌐ [GRAPH].

Note that there is only one point of intersection for the two functions, the market equilibrium point with input x ≈ 1.8331. Recall that we stored the complete value of x in E.

Total social gain = producers' surplus + consumers' surplus at the equilibrium price. Shade this area on your graph and then use the graph to write the integrals that give this area:

$$\text{Social gain} = \int_1^E S(p)\,dp + \int_E^\infty D(p)\,dp \text{ where E} \approx 1.8331$$

Find the producer's surplus as shown to the right.

The TI-86 does not calculate the value of improper integrals. However, as a check on your algebraic work to determine the value of the consumer's surplus, you can do the following.

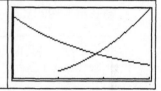

Enter fnInt(y1, x, E, x) in a location of the y(x)= list, say y3. Turn off Y1 and y2. Set the TI-86 TABLE to ASK and enter increasingly larger values of x such as those shown to the right.

It appears that $\lim\limits_{N\to\infty} \int_E^N D(p)\,dp \approx 4.360$, giving social gain ≈ 1.108 + 4.360 ≈ $5.5 million.

NOTE: The remainder of the material in this *Guide* refers to the complete text for *Calculus Concepts* (that also contains Section 7.4 and Chapters 8-11.)

 ## 7.4 Probability Distributions and Density Functions

Most of the applications of probability distributions and density functions use technology techniques that have already been discussed. Probabilities are areas whose values can be found by integrating the appropriate density function. A cumulative density function is an accumulation function of a probability density function.

Your calculator's numerical integrator is especially useful for finding means and standard deviations of some probability distributions because those integrals often involve expressions for which we have not developed algebraic techniques for finding antiderivatives.

7.4.1a NORMAL PROBABILITIES   The normal density function is the most well known and widely used probability distribution. If you are told that a random variable x has a normal distribution $N(x)$ with mean μ and standard deviation σ, the probability that x is between two real numbers a and b is given by

$$\int_a^b N(x)\,dx = \int_a^b \frac{1}{\sigma\sqrt{2\pi}}\, e^{\frac{-(x-\mu)^2}{2\sigma^2}}\, dx$$

We illustrate these ideas with the situation in Example 8 of Section 7.4 of *Calculus Concepts*. In that example, we are given that the distribution of the life of the bulbs, with the life span measured in hundreds of hours, is modeled by a normal density function with $\mu = 900$ hours and $\sigma = 100$ hours. Carefully use parentheses when entering the normal density function.

The probability that a light bulb lasts between 9 and 10 hundred hours is $$P(9 \leq x \leq 10) = \int_9^{10} \frac{1}{\sqrt{2\pi}}\, e^{-(x-9)^2/2}\, dx$$ The value of this integral must be a number between 0 and 1.	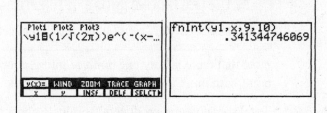

7.4.1b VIEWING NORMAL PROBABILITIES   We could have found the probability that the light bulb life is between 900 and 1000 hours graphically by using the CALC menu.

To draw a graph of the normal density function, note that just about all of the area between this function and the horizontal axis lies within three standard deviations of the mean. Set xMin = 9 − 3 = 6 and xMax = 9 + 3 = 12. Use ZFIT to draw the graph. Reset yMin to ⁻0.1 in preparation for the next step, and press F5 [GRAPH].	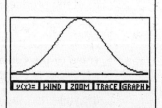
With the graph on the screen, press MORE F1 [MATH] F3 [∫f(x)]. At the Lower Limit? prompt, enter 9. At the Upper Limit? prompt, enter 10. The probability is displayed as an area and the value is printed at the bottom of the graphics screen.	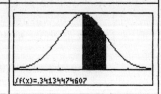

Chapter 8   Repetitive Change:
Cycles and Trigonometry

 ## 8.1 Functions of Angles: Sine and Cosine

Before you begin this chapter, go back to the first page of the *Graphing Calculator Instruction Guide* and check the basic setup, the statistical setup, and the window setup. If these are not set as specified in Figures 1, 2, and 3, you will have trouble using your calculator in this chapter. Pay careful attention to the third line in the MODE screen in the basic setup. The Radian/Degree mode setting affects the TI-86's interpretation of the ANGLE menu choices. The calculator's MODE menu should always be set to Radian unless otherwise specified. (Note that calculator instructions for material that is in *Appendix A*: *Trigonometry Basics* are on the *Calculus Concepts* CD-ROM and web site.)

8.1.1 FINDING OUTPUTS OF TRIG FUNCTIONS WITH RADIAN INPUTS   It is essential that you have the correct mode set when evaluating trigonometric function outputs. The angle setting in the MODE menu must be Radian for all applications in Chapter 8. We illustrate how to evaluate trig functions with the following example.

Find $\sin \frac{9\pi}{8}$ and $\cos \frac{9\pi}{8}$. Because these angles are in radians, be certain that Radian is chosen in the third line of the MODE screen. The $\boxed{\text{SIN}}$ and $\boxed{\text{COS}}$ keys are above the $\boxed{\text{EE}}$ and $\boxed{(}$ keys on the TI-86 keyboard.	`sin (9π/8)`
	`          -.382683432365`
	`cos (9π/8)`
	`          -.923879532511`

It is also essential that you use parentheses – the TI-86's order of operations is such that the sine function takes precedence over division. Unless you tell the TI-86 to first divide by using parentheses around $\frac{9\pi}{8}$, it evaluates $(\sin 9) \cdot \frac{\pi}{8}$ and $(\cos 9) \cdot \frac{\pi}{8}$.	`                  -.382683432365`
	`cos (9π/8)`
	`          -.923879532511`
	`sin 9π/8  .161838550706`
	`cos 9π/8`
	`          -.35780001715`

 ## 8.2 Cyclic Functions as Models

We now introduce another model – the sine model. As you might expect, this function should be fit to data that repeatedly varies between alternate extremes. The form of the sine model is given by $f(x) = a \sin (bx + h) + k$ where $|a|$ is the amplitude, b is the frequency (where $b > 0$), $2\pi/b$ is the period, $|h|/b$ is the horizontal shift (to the right if $h < 0$ and to the left if $h > 0$), and k is the vertical shift (up if $k > 0$ and down if $k < 0$). *Note:* The TI-86 uses the c when we use h and d when we use k.

8.2.1 FITTING A SINE MODEL TO DATA   Before fitting any model to data, remember that you should construct a scatter plot of the data and observe what pattern the data appear to follow. Example 2 in Section 8.2 asks you to find a sine model for cyclic data with the hours of daylight on the Arctic Circle as a function of the day of the year on which the hours of daylight are measured. (January 1 is day 1.) These data appear in Table 8.3 of *Calculus Concepts*.

Day of the year	−10	81.5	173	264	355	446.5	538	629	720	811.5
Hours of daylight	0	12	24	12	0	12	24	12	0	12

Clear any old data. Delete any functions in the y(x)= list and turn on Plot 1. Enter the data in the above table in lists L1 and L2. Construct a scatter plot of the data. When using the sine regression in the TI-86, it is sometimes necessary to have an estimate of the period of the data.

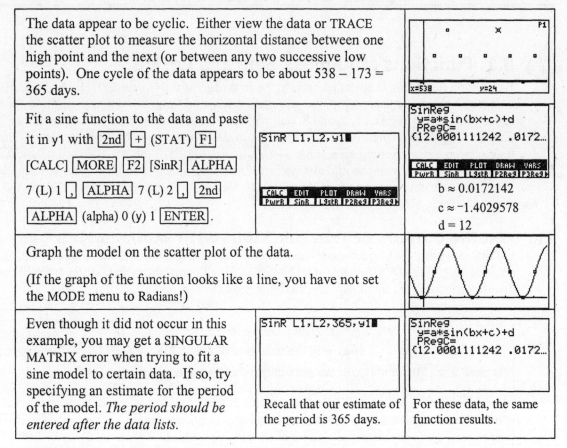

The data appear to be cyclic. Either view the data or TRACE the scatter plot to measure the horizontal distance between one high point and the next (or between any two successive low points). One cycle of the data appears to be about $538 - 173 = 365$ days.

Fit a sine function to the data and paste it in y1 with [2nd] [+] (STAT) [F1] [CALC] [MORE] [F2] [SinR] [ALPHA] 7 (L) 1 [,] [ALPHA] 7 (L) 2 [,] [2nd] [ALPHA] (alpha) 0 (y) 1 [ENTER].

$b \approx 0.0172142$

$c \approx -1.4029578$

$d = 12$

Graph the model on the scatter plot of the data.

(If the graph of the function looks like a line, you have not set the MODE menu to Radians!)

Even though it did not occur in this example, you may get a SINGULAR MATRIX error when trying to fit a sine model to certain data. If so, try specifying an estimate for the period of the model. *The period should be entered after the data lists.*

Recall that our estimate of the period is 365 days.

For these data, the same function results.

NOTE: If you do not think the original function the calculator finds fits the data very well, try specifying a period and see if a better-fitting equation results. It didn't here, but it might with a different set of data. If you still cannot find an equation, you can tell the TI-86 how many times to go through the routine that finds the equation. This number of iterations is 3 if not specified. The number should be typed before L1 when initially finding the equation.

8.3 Rates of Change and Derivatives

All the previous techniques given for other functions also hold for the sine model. You can find intersections, maxima, minima, inflection points, derivatives, integrals, and so forth.

8.3.1 DERIVATIVES OF SINE AND COSINE FUNCTIONS Evaluate der1 at a particular input to find the value of the derivative of the sine function at that input. We illustrate with Example 1 in Section 8.3 of *Calculus Concepts*:

> The calls for service made to a county sheriff's department in a certain rural/suburban county can be modeled as $c(t) = 2.8 \sin(0.262t + 2.5) + 5.38$ calls during the *t*th hour after midnight.

Part *a* of this example asks for the average number of calls the county sheriff's department receives each hour. The easiest way to obtain this answer is to remember that the parameter k (d in the TI-86) in the sine function is the average value.

You can also find the average value over one period of the function using the methods of Section 6.6.1a of the *Graphing Calculator Instruction Guide*. Enter the function c in y1 and type in the quotient shown to the right.	Average Value = $$\frac{\int_0^{2\pi/0.262} c(t)\,dt}{2\pi/0.262 - 0}$$	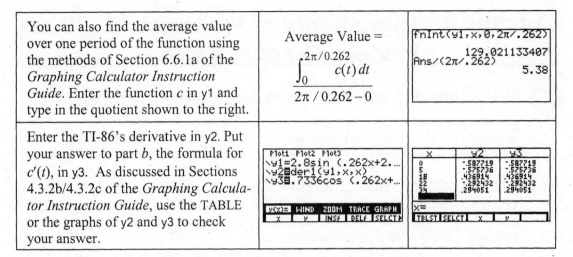 `fnInt(y1,x,0,2π/.262)` `129.021133407` `Ans/(2π/.262)` `5.38`
Enter the TI-86's derivative in y2. Put your answer to part b, the formula for $c'(t)$, in y3. As discussed in Sections 4.3.2b/4.3.2c of the *Graphing Calculator Instruction Guide*, use the TABLE or the graphs of y2 and y3 to check your answer.	Plot1 Plot2 Plot3 \y1=2.8sin (.262x+2... \y2=der1(y1,x,x) \y3=.7336cos (.262x+...	x / y2 / y3 table

We entered $c'(x) = 0.7336 \cos(0.262x + 2.5)$ in y3. Your formula may be different. We now go to part c of Example 1 that asks how quickly the number of calls received each hour is changing at noon and at midnight.

To answer these questions, simply evaluate y2 (or your derivative in y3) at 12 for noon and 0 (or 24) for midnight. (You were not told if "midnight" refers to the initial time or 24 hours after that initial time.)	`y2(12)` `.588774171694` `y2(0)` `-.587718956365` `y2(24)` `-.589825974892`

8.5 Accumulation in Cycles

As with the other functions we have studied, applications of accumulated change with the sine and cosine functions involve the calculator's numerical integrator fnInt.

8.5.1 INTEGRALS OF SINE AND COSINE FUNCTIONS We illustrate the process of determining accumulated change with Example 1 in Section 8.5 of *Calculus Concepts* that uses the rate of change of temperature in Philadelphia on August 27, 1993: $t(x) = 2.733 \cos(0.285x - 2.93)$ °F per hour, x hours after midnight.

Enter the function t in y1. Find the accumulated change in the temperature between 9 a.m. and 3 p.m. using fnInt.	Plot1 Plot2 Plot3 \y1=.733cos (.285x-...	`fnInt(y1,x,9,15)` `12.7690116202`

The temperature increased by about 13°F between 9 a.m. and 3 p.m.

Chapter 9   Ingredients of Multivariable Change: Models, Graphs, Rates

 ## 9.1 Multivariable Functions and Contour Graphs

Because any program that we might use to graph[1] a three-dimensional function would be fairly involved and take a long time to execute, we do not graph three-dimensional functions. Instead, we discover information about three-dimensional graphs using their associated contour curves.

9.1.1 SKETCHING CONTOUR CURVES   When given a multivariable function with two input variables, you can draw contour graphs using the three-step process described below. We illustrate with a function that gives body heat loss due to the wind. This function, H, appears in Example 2 of Section 9.1 of *Calculus Concepts*:

$$H(v, t) = (10.45 + 10\sqrt{v} - v)(33 - t) \text{ kilogram-calories}$$

per square meter of body surface area per hour for wind speed v in meters per second.

Step 1:   Set $H(v, t) = 2000$. We will use the SOLVER to find the value of t at various values of v that satisfy the equation $H(v, t) = 2000$.

Step 2:   Choose values for v and solve for t to obtain points on the 2000 constant-contour curve. Obtain guesses for the values of v and t from Table 9.2 in the text.

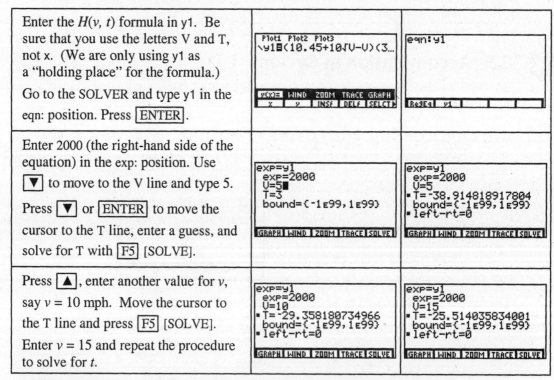

WARNING: The cursor must be on the line corresponding to the unknown variable for the SOLVER to solve the equation for that variable. If you do not have a table of values for the quantities, you should enter several different guesses for the unknown variable to determine whether there is more than one solution to the equation.

---

[1] There are many TI-86 programs, including one that will graph a multivariable function with two input variables, available at the Texas Instruments web site with address http://education.ti.com.

| Repeat the procedure for $v = 20$ and $v = 25$. Make a table of the values of v and t as you find them. | exp=y1
exp=2000
v=20
•T=-23.864449529083
bound={-1E99,1E99}
•left-rt=0
GRAPH WIND ZOOM TRACE SOLVE | exp=y1
exp=2000
v=25
•T=-23.417489421721
bound={-1E99,1E99}
•left-rt=0
GRAPH WIND ZOOM TRACE SOLVE |

Step 3: Plot the points obtained in Step 2 with pencil and paper. You need to find as many points as it takes to see the pattern the points are indicating when you plot them. Connect the points you have plotted with a smooth curve.

You must have a function given to draw a contour graph using the above method. Even though there may be several functions that seem to fit the data points obtained in Step 2, their use would be misleading because the real best-fit function can only be determined by substituting the appropriate values in a multivariable function. The focus of this section is to use contour graphs to study the relationships between input variables, not to find the equation of a function to fit a contour curve. Thus, we always sketch the contours on paper rather than with the TI-86.

9.2 Cross-Sectional Models and Rates of Change

For a multivariable function with two input variables, obtain a cross-sectional model by entering the data in lists L1 and L2 and then fitting the appropriate function as indicated in previous chapters of this *Guide*. Unless you are told otherwise, we assume that the data are given in a table with the values of the *first* input variable listed *horizontally* across the top of the table and the values of the *second* input variable listed *vertically* down the left side of the table.

9.2.1a FINDING A CROSS-SECTIONAL MODEL FROM DATA (HOLDING THE FIRST INPUT VARIABLE CONSTANT)

We are to find the cross-sectional model $E(0.8, n)$ using the elevation data that appear in Table 9.4 of Section 9.2 in *Calculus Concepts*. However, the data you use in the activities will be obtained from a multivariable table. So we digress for a moment to understand how to find the data you need in such a table.

Refer to Table 9.3 on page 622 of the text. Remember that "rows" (containing the n data) go from left to right horizontally and "columns" (containing the e data) go from top to bottom vertically. In $E(0.8, n)$, e is constant at 0.8 and n varies. Thus, choose the values for n that appear on the left side of the table (vertically) in L1 and the elevations E in the $e = 0.8$ column of Table 9.3 in L2. *In general, the inputs that you enter in L1 are either across the top or down the left side of the table, and the outputs that you enter in L2 will always be in the main body of any multivariable data table in this text.* Thus the data we enter are these values:

n (miles)	1.5	1.4	1.3	1.2	1.1	1.0	0.9	0.8
Elevation (feet above sea level)	797.6	798.1	798.5	798.9	799.2	799.5	799.7	799.9
n (miles)	0.7	0.6	0.5	0.4	0.3	0.2	0.1	0.0
Elevation (feet above sea level)	800.0	800.1	800.1	800.1	800.0	799.9	799.7	799.5

| After entering the data, clear any previously entered functions from the y(x)= list, and turn on Plot1. Draw a scatter plot of the data with [F3] [ZOOM] [MORE] [F5] [ZDATA]. | 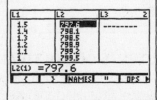 | 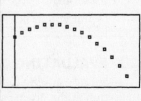 |

The data appear to be quadratic. Fit a quadratic function and copy it to the y(x)= list. (Refer to Section 2.4.1b of this *Guide*.) Overdraw the graph of the function on the scatter plot with GRAPH F5 [GRAPH].

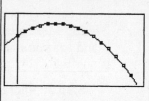

CAUTION: Because you will often be asked to find several different cross-sectional models using the same data table, calling different variables by the same names *x* and *y* would be very confusing. It is very important that you call the variables by the names that have been assigned in the problem. Remember that when finding or graphing a function, the TI-86 always calls the input variable x and the output variable y. When working with multivariable functions, you must translate the calculator's equation $y1 \approx -2.5x^2 + 2.497x + 799.490$ into the symbols that are used in the application. You should write the cross-sectional function as $E(0.8, n) = -2.5n^2 + 2.497n + 799.490$. Don't forget to completely describe (including units) all of the variables.

9.2.1b FINDING A CROSS-SECTIONAL MODEL FROM DATA (HOLDING THE SECOND INPUT VARIABLE CONSTANT)

The only difference in this model and the one in the previous section of this *Guide* is that the second input, instead of the first, is held constant. Refer again to Table 9.3 on page 622 of the text. Because we are asked to find the cross-sectional model $E(e, 0.6)$, $n = 0.6$ and the inputs are the values of *e* that are across the top of the table. Enter these values in L1. (See the first box below for a shortcut.) The outputs are the elevations *E* obtained in the $n = 0.6$ mile row in Table 9.3. Enter these outputs in L2.

NOTE: You may find it helpful to place a piece of paper or a ruler under the row (or to the right of the column) in which the data appear to help avoid entering an incorrect value.

Because the input values for this function are the same as the input values in the last section of this *Guide* (but in reverse order) you can sort L1 in ascending order to avoid re-entering the data. On the home screen, press 2nd – (LIST) F5 [OPS] F2 [sortA] ALPHA 7 (L) 1 ENTER. Store the sorted list in L1 and then enter the output in L2.

After entering the data, clear any functions from the y(x)= list, and turn on Plot1. Draw a scatter plot of the data. (See Section 1.4.2b.)

There is an inflection point and no evidence of limiting values, so the data appear to be cubic. Fit a cubic function and copy it to the y(x)= list. (Refer to Section 2.4.2 of this *Guide*.) Draw the cubic function on the scatter plot with GRAPH.

9.2.2 EVALUATING OUTPUTS OF MULTIVARIABLE FUNCTIONS

As is the case with single-variable functions, outputs of multivariable functions are found by evaluating the function at the given values of the input variables. The main difference is that you usually will not be using x as the input variable symbol. One way to find multivariable function outputs is to evaluate them on the home screen. We illustrate with the investment function in Example 1 of Section 9.2 in *Calculus Concepts*.

The answer to part *a* of Example 1, as derived from the compound interest formula, uses the formula for the accumulated amount of an investment of *P* dollars for *t* years in an account paying 6% interest compounded quarterly:

$$A(P, t) = P\left(1 + \frac{0.06}{4}\right)^{4t} \text{ dollars}$$

When 10 is substituted for *t*, the cross-sectional function becomes $A(P, 10) \approx 1.814018409P$. Part *b* of Example 1 asks for $A(5300, 10)$. Even though it is simplest here to substitute 5300 for *P* in $A(P, 10) \approx 1.814018409P$, we return to the original function to illustrate evaluating multivariable formulas on the calculator.

To find the output on the home screen, type the formula for $A(P, t)$, substituting $P = 5300$ and $t = 10$. Press ENTER. Again, be warned that you must carefully use the correct placement of parentheses.	`5300(1+.06/4)^(4*10)` `          9614.29756594`

Even though it is not necessary in this example, you may encounter activities in this section in which you need to evaluate a multivariable function at several different inputs. You could use what is shown above, but there are easier methods than individually entering each calculation. You will also use the techniques shown below in later sections of this chapter. When evaluating a multivariable function at several different input values, you may find it more convenient to enter the multivariable function in the graphing list.

Clear any previously entered equations. Enter in y1 the function $A(P, t) = P\left(1 + \frac{0.06}{4}\right)^{4t}$ with the keystrokes ALPHA $\square$ (P) $\square$ 1 $\boxed{+}$ $\square$ 06 $\boxed{\div}$ 4 $\boxed{)}$ $\boxed{\wedge}$ $\square$ 4 ALPHA $\boxed{-}$ (T) $\boxed{)}$.	`Plot1 Plot2 Plot3`
	 `\y1⊟P(1+.06/4)^(4T)`
	
	 `y(x)= WIND ZOOM TRACE GRAPH`
	 `  x   y   INSf DELf SELCT▶`

Return to the home screen and input the values $P = 5300$ and $t = 10$ with the keystrokes 5300 STO▶ $\square$ (P) 2nd $\boxed{.}$ (:) ALPHA 10 STO▶ $\boxed{-}$ (T) ENTER. Evaluate y1 at these inputs with 2nd ALPHA (alpha) 0 (y) 1 ENTER.	`5300→P:10→T`
	 `                    10`
	 `y1`
	 `          9614.29756594`

To evaluate y1 at other inputs, press 2nd ENTER (ENTRY) twice to recall the storing instruction. Change the values and press ENTER. Then press 2nd ENTER (ENTRY) twice to recall y1 and press ENTER.	`y1                  10`
	 `          9614.29756594`
	 `6500→P:8→T`
	 `                     8`
	 `y1`
	 `          10467.108081`

CAUTION: It is very important to note at this point that while we have previously used x as the input variable when entering functions in the y(x)= list, we do *not* follow this rule when we *evaluate* functions with more than one input variable. However, realize that we should not graph y1 nor use the TABLE. If you attempt to graph the current $y1 = A(P, t)$ or use the table, you will see that the calculator considers this y1 a constant. (Check it out and see that the graph is a horizontal line at about 10467.1 and that all values in the table are about 10467.1.)

9.2.3 VISUALIZING AND ESTIMATING RATES OF CHANGE OF CROSS SECTIONS

The rate of change of a multivariable function (when evaluated at a specific point) is the slope of the line tangent to the graph of a cross-sectional function at that point. We illustrate this concept in this section and the next using the Missouri farmland cross-section equations for elevation: $E(0.8, n)$ and $E(e, 0.6)$. It would be best to use the unrounded functions that were found in Sections 9.2.1a and 9.2.1b of this *Guide*. However, to illustrate the rate-of-change techniques, we use the rounded functions rather than re-enter all the data.

Enter $-2.5n^2 + 2.497n + 799.490 =$ $E(0.8, n)$ in y1. Because we are going to graph this function, use x, not n, as the input variable. Press EXIT F2 [WIND] and set values such as those shown to the right.	Plot1 Plot2 Plot3 \y1■-2.5x²+2.497x+79… y(x)= WIND ZOOM TRACE GRAPH x | y | INSf | DELf | SELCTf	WINDOW xMin=0 xMax=1.5 xScl=1 yMin=795 yMax=804 ↓yScl=1 y(x)= | WIND | ZOOM | TRACE | GRAPH▶

The window settings used above can be obtained by drawing a scatter plot of the data used to find $E(0.8, n)$ or by looking at the $e = 0.8$ column in Table 9.3 in *Calculus Concepts*. The line tangent to the graph at $n = x = 0.6$ can be drawn from either the home screen or the graphics screen. (See Section 3.2.1b of this *Guide* for an explanation of both methods.) We use the home screen method in this section and the graphics screen method in the next.

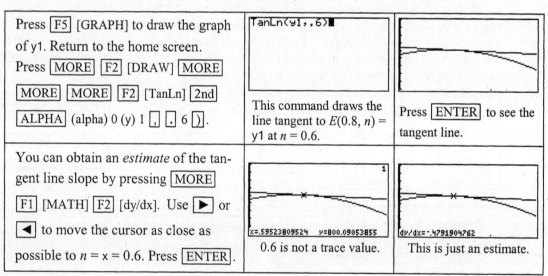

Press F5 [GRAPH] to draw the graph of y1. Return to the home screen. Press MORE F2 [DRAW] MORE MORE MORE F2 [TanLn] 2nd ALPHA (alpha) 0 (y) 1 ⌨, ⌨. 6 ⌨).	This command draws the line tangent to $E(0.8, n) =$ y1 at $n = 0.6$.	Press ENTER to see the tangent line.
You can obtain an *estimate* of the tangent line slope by pressing MORE F1 [MATH] F2 [dy/dx]. Use ▶ or ◀ to move the cursor as close as possible to $n = x = 0.6$. Press ENTER.	x=.59523809524 y=800.09053855 0.6 is not a trace value.	dy/dx=-.4791904762 This is just an estimate.

We could have obtained a much better estimate than the one given above if we had found the slope of the tangent line at x = 0.6 rather than a value close to that number. The TI-86 allows you to do this on the home screen or graphically. Here we illustrate the graphical method.

Have $E(0.8, n)$, using x as the input variable, in y1. Draw the tangent line to $E(0.8, n)$ at $n = 0.6$ by first drawing the graph of the function with F5 [GRAPH]. On the graph screen, press MORE F1 [MATH] MORE MORE F1 [TANLN] ⌨. 6 ENTER.	dy/dx=-.503

The slope calculated by der1 is shown. So, $\dfrac{dE(0.8, n)}{dn}$ at $n = 0.6$ is -0.503 foot/mile.

9.2.4 FINDING RATES OF CHANGE USING CROSS-SECTIONAL MODELS The methods described in the last section require that you use x as the input variable to find rates of change. The process of replacing other variables with x can be confusing. You can avoid having to do this replacement when you use the TI-86 numerical derivative on the home screen to evaluate rates of change at specific input values. We illustrate with the cross-sectional function $E(e, 0.6)$ $= -10.124e^3 + 21.347e^2 - 13.972e + 802.809$ feet above sea level that was found in Section 9.2.1b of this *Guide*. We also demonstrate this method with the cross-section $E(0.8, n)$ that was found in Section 9.2.1a and used in the last section of this *Guide*.

NOTE: Whenever possible, you should use the complete equation found by the TI-86 with this method. However, to avoid re-entering the data, we use the rounded functions to illustrate.

Have $E(0.8, n) = -2.5n^2 + 2.497n + 799.490$, using N as the input variable, in y1. Enter $E(e, 0.6)$, using E as the input variable, in y2. (Remember that because we are not using x as the input variable, we should not draw a graph or use the TABLE.)	

NOTE: These functions could be entered on the home screen instead of in the y(x)= list. We use y1 and y2 in the y(x)= list because they are convenient locations to hold the equations.

Tell the TI-86 the point at which the derivative is to be evaluated by storing the values. Then, on the home screen, find $\dfrac{dE(0.8, n)}{dn}$ evaluated at $n = 0.6$ with the derivative der1.	
Find $\dfrac{dE(e, 0.6)}{de}$ evaluated at $e = 0.8$ to be about 0.745 foot per mile. (Always remember to attach units of measure to the numerical values when writing your answer.)	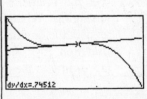

NOTE: It is not necessary to again store the values for N and E unless for some reason they have been changed from the values stored previously.

If you want to see the line tangent to the graph of $E(e, 0.6)$ at $e = 0.8$, first change every E in y2 to x and turn off y1. Draw the graph of y2 and use the TANLN instruction in the DRAW MATH menu.	

9.3 Partial Rates of Change

When holding all but one of the input variables in a multivariable function constant, you are actually looking at a function of one input variable. Thus, all of the techniques for finding derivatives that we discussed previously can be used. In particular, the calculator's numerical derivatives, der1 and nDer, can be used to find partial rates of change at specific values of the varying input variable.

Although your TI-86 does not give formulas for derivatives, you can use it as discussed in Sections 4.3.2b and/or 4.3.2c of this *Guide* to check your answer for the algebraic formula for a partial derivative.

9.3.1 NUMERICALLY CHECKING PARTIAL DERIVATIVE FORMULAS   As mentioned in Chapter 4, the basic concept in checking your algebraically-found partial derivative formula is that your formula and the calculator's formula computed with der1 or nDer should have the same outputs when each is evaluated at several different randomly-chosen inputs. You can use the methods in Section 9.2.2 of this *Guide* to evaluate each derivative formula at several different inputs and determine if the same numerical values are obtained from each formula.

We illustrate these ideas by checking the answers for the partial derivative formulas found in parts *b* and *d* of Example 1 in Section 9.3 of *Calculus Concepts* for the following function:

The accumulated value of an investment of *P* dollars over *t* years at an APR of 6% compounded quarterly is $A(P, t) = P(1.061363551^t)$.

Recall that the syntax for the calculator's derivative that we use whenever possible is

der1(function, symbol for input variable, point at which the derivative is evaluated)

Enter the function *A*, using the letters P and T that appear in the formula, in y1. Part *b* of Example 1 asks for a formula for $\partial A/\partial t$, so enter your formula (which may not be the same as the one that is shown to the right) in y2. Enter the TI-86's derivative, using T as the changing input, in y3.	``` Plot1 Plot2 Plot3 \y1▣P(1.061363551^T) \y2▣P ln(1.061363551… \y3▣der1(y1,T,T) x y INSf DELf SELCT evalF nDer der1 der2 fnInt ▶ ```

Store a value in T and call up y2 and y3. Do this for some different inputs. If the outputs of y2 and y3 are the same, your answer in y2 is *probably* correct.	```
	5→T:y2
	         521.372863237
	y3
	         521.372863237
	10→T:y2
	         702.21365237
	▪
	``` ```
	y3 702.21365237
	       702.21365237
	25→T:y2
	       1715.66229132
	y3
	       1715.66229132
	``

NOTE: Remember that [2nd] [ENTER] (ENTRY) recalls previously entered statements so you do not have to spend time re-entering them. Also recall that [2nd] [.] (:) joins two or more statements together. Only the value of the last statement is printed, so do not join all three statements shown in the box above together with this symbol.

You may find it more convenient to numerically check your answer using the TABLE. If so, you must remember that when using the TABLE, the TI-86 considers x as the variable that is changing. When finding a partial derivative formula, all other variables are held constant except the one that is changing. So, to use the TABLE (or draw a graph) with a multivariable function, just store values in all constants and call the changing variable x. Then, proceed according to the directions given in Chapter 4. We illustrate this with part *d* of Example 1.

Have the function *A*, using the letters P and T that appear in the formula, in y1. Because P is the variable that is changing in $\partial A/\partial P$, replace P with x. Turn off y1. Enter your formula for $\partial A/\partial P$ (which may not be the same as the one that is shown to the right) in y2. Enter the TI-86's derivative, using x as the changing input, in y3.	``` Plot1 Plot2 Plot3 \y1=x(1.061363551^T) \y2▣(1.061363551^T) \y3▣der1(y1,x,x) y(x)= WIND ZOOM TRACE GRAPH x y INSf DELf SELCT▶ ```
Return to the home screen. Because T is constant, enter any reasonable value, except 0, in T. (You could repeat what comes next for T = 2, 10, and 37 as shown in part *c* of Example 1.) (If you need additional instructions on using the TABLE, see Section 1.1.1f of this *Guide*.)	``` 2→T 2 ```

Have TBLSET set to ASK in the Indpnt: location. Go to the TABLE and input some reasonable values for $t = x$. If the values shown for y2 and y3 are the same, your formula in y2 for $\partial A/\partial P$ is *probably* correct.	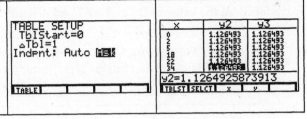

Why are the values in y2 and y3 the same for all values of T? It is because the function in y1 is a linear function with a constant slope. Remember that T is a constant because P is the changing variable in $\partial A/\partial P$.

NOTE: If you have an unrounded function (one fit to data) in y1 and you use the rounded function to compute your partial derivative, the y2 and y3 columns will be slightly different.

9.4 Compensating for Change

As you have just seen, the TI-86 closely estimates numerical values of partial derivatives with nDer and finds the numerical value of the derivative with its der1 function. This technique can also be very beneficial and help you eliminate many potential calculation mistakes when you find the rate of change of one input variable with respect to another input variable (that is, the slope of the tangent line) at a point on a contour curve.

9.4.1a EVALUATING PARTIAL DERIVATIVES OF MULTIVARIABLE FUNCTIONS The last few sections of this *Guide* indicate how to estimate and evaluate partial derivatives using cross-sectional models. The TI-86 evaluates partial derivatives calculated directly from multi-variable function formulas using the same procedures. The most important thing to remember is that you must supply the name of the input variable that is changing and the values at which the partial derivative is evaluated. We illustrate using the body-mass index function that is in Example 1 of Section 9.4 of *Calculus Concepts*:

A person's body-mass index is given by $B(h, w) = \dfrac{0.4536w}{0.00064516h^2}$ where h is the person's height in inches and w is the person's weight in pounds. We first find B_h and B_w at a specific height and weight and then use those values in the next section of this *Guide* to find the value of the derivative $\dfrac{dw}{dh}$ at that particular height and weight. The person in this example is 5 feet 7 inches tall and weighs 129 pounds.

Enter B in the y1 location of the y(x)= list, using the letters H and W for the input variables. Next, store the values of H and W at the given point.	Plot1 Plot2 Plot3
	\y1∎.4536W/(.0006451…
	
	y(x)= WIND ZOOM TRACE GRAPH
	x y INSf DELf SELCT▶
		67→H
		          67
		129→W
		         129

CAUTION: Forgetting to store the values of the point at which the derivative is to be evaluated is a common mistake made when using this method. The TI-86 derivative instructions specify the value at only one input, and you must tell the calculator the values of all inputs.

Find the value of B_h at $h = 67$ and $w = 129$ by evaluating the expression shown to the right. The symbol B_h (or $\partial B/\partial h$) tells you that h is varying and w is constant, so h is the variable you type in the numerical derivative. Next, enter the value of h. Store this result in N for use in the next section of this *Guide*.	der1(y1,H,67)
	      -.60311608406
	Ans→N
	      -.60311608406
	
	evalF nDer der1 der2 fnInt▶

> Find the value of B_w at $h = 67$ and $w = 129$ by evaluating the expression shown to the right. The symbol B_w (or $\partial B/\partial w$) tells you that w is varying and h is constant, so w is the variable you type in the numerical derivative. Next, enter the value of w.
>
> Store this result in D for use in the next section of this *Guide*.

```
der1(y1,W,129)
            .156623169116
Ans→D
            .156623169116


evalF nDer  der1  der2 fnInt
```

9.4.1b FINDING THE SLOPE OF A LINE TANGENT TO A CONTOUR CURVE

We continue the previous illustration with the body-mass index function in Example 1 of Section 9.4 of *Calculus Concepts*. Part *a* of Example 1 asks for $\frac{dw}{dh}$ at the point (67, 129) on the contour curve corresponding to the person's current body-mass index. The formula is $\frac{dw}{dh} = \frac{-B_h}{B_w}$.

An easy way to remember this formula is that whatever variable is in the numerator of the derivative (in this case, w) is the same variable that appears as the changing variable in the denominator of the slope formula. This is why we stored B_w as D (for denominator) and B_h as N (for numerator). Don't forget to put a minus sign in front of the numerator.

> In the previous section, we stored B_h as N and B_w as D. So,
>
> $\frac{dw}{dh} = \frac{-B_h}{B_w} = -N \div D$. (Storing these values also avoids round-off error.) The rate of change is about 3.85 pounds per inch.

```
-N/D
            3.85074626866
```

9.4.1c COMPENSATING FOR CHANGE

When one input of a two-variable multivariable function changes by a small amount, the value of the function is no longer the same as it was before the change. The methods illustrated below show how to determine the amount by which the other input must change so that the output of the function remains at the value it was before any changes were made. We again continue the previous illustration with the body-mass index function and part *b* of Example 1 of Section 9.4 of *Calculus Concepts*.

> To estimate the change in weight needed to compensate for growths of 0.5 inch, 1 inch, and 2 inches if the person's body-mass index is to remain constant, you need to find
>
> $$\Delta w \approx \frac{dw}{dh}(\Delta h)$$
>
> at the given values of Δh.
>
> Again, to avoid rounding error, it is easiest to store the slope of the tangent line to some location, say T, and use the unrounded value to calculate.

```
                3.85074626866
Ans→T
                3.85074626866
T*.5
                1.92537313433
T*2
                7.70149253731
```

These are changes in weight, so the units that should be attached are *pounds*.

Chapter 10   Analyzing Multivariable Change: Optimization

 ## 10.2 Multivariable Optimization

As you might expect, multivariable optimization techniques that you use with your TI-86 are very similar to those that were discussed in Chapter 5. The basic difference is that the algebra required to get the expression that comes from solving a system of equations with several unknowns reduced to one equation in one unknown is sometimes difficult. However, once your equation is of that form, all the optimization procedures are basically the same as those that were discussed previously.

10.2.1a FINDING CRITICAL POINTS USING ALGEBRA AND THE SOLVER   Critical points for a multivariable function are points at which maxima, minima, or saddle points occur. We begin the process of finding critical points of a smooth, continuous mutivariable function by using derivative formulas to find the partial derivative with respect to each input variable and setting these partial derivatives each equal to 0. We next use algebraic methods to obtain one equation in one unknown input. Then, you can use your calculator's solver to obtain the solution to that equation. This method of solution works for all types of equations. We illustrate these ideas with the postal rate function given at the beginning of Section 10.2:

> The volume of a rectangular package that contains the maximum amount of printed material and is sent at the bound printed matter postal rate is given by
>
> $V(h, w) = 108hw - 2h^2w - 2hw^2$ cubic inches
>
> where h inches is the height and w inches is the width of the package.

Find the two partial derivatives and set each of them equal to 0 to obtain these equations:

$$V_h: \quad 108w - 4hw - 2w^2 = 0 \qquad\qquad [1]$$
$$V_w: \quad 108h - 2h^2 - 4hw = 0 \qquad\qquad [2]$$

WARNING: Everything that you do with your calculator depends on the partial derivative formulas that you find using derivative rules. Be certain that you check your work before using any of the following solution methods.

Next, solve one of the equations for one of the variables, say equation 2 for w, to obtain

$$w = \frac{108h - 2h^2}{4h}$$

Remember that to *solve* for a quantity means that it must be by itself on one side of the equation without the other side of the equation containing that letter. Now, let your TI-86 work.

Clear the y(x)= list, and enter the function V in y1. Remember that you must use ⊗ between two variables that are written next to each other to indicate that they are to be multiplied. Enter the expression for w in y2. Be certain that you enclose numerators and denominators of fractions in parentheses.	Plot1 Plot2 Plot3 \y1≣108H∗W-2H²∗W-2H∗… \y2≣(108H-2H²)/(4H) \y3= y(x)= WIND ZOOM TRACE GRAPH x y INSf DELf SELCT► This step tells the TI-86 that $w = $ y2.
Type the left-hand side of the *other* equation (here, equation 1) in y3. Equations 1 and 2 are now in the TI-86. Remember that when using this method, the expression you type in y3 must be equal to zero.	Plot1 Plot2 Plot3 \y1≣108H∗W-2H²∗W-2H∗… \y2≣(108H-2H²)/(4H) \y3≣108W-4H∗W-2W² y(x)= WIND ZOOM TRACE GRAPH x y INSf DELf SELCT►

Now, replace *every* W in y3 with the symbol y2 by placing the cursor on each W location and pressing 2nd ALPHA (alpha) 0 (Y) 1 DEL (INS) 2. What you have just done is substitute the expression for W from equation 2 in equation 1!

```
Plot1 Plot2 Plot3
\y1⊟108H*W-2H²*W-2H*…
\y2⊟(108H-2H²)/(4H)
\y3⊟108y2-4H*y2-2y2²
────────────────────
x(x)² WIND ZOOM TRACE GRAPH
  x   y   INSf DELf SELCT▶
```

NOTE: The expression in y3 is the left-hand side of an equation that equals 0 and it contains only one variable, namely *h*.

The next step is to use the SOLVER to solve the equation y3 = 0. Try different guesses and see that they all result in the same solution for this particular equation.

```
eqn:y3▮



────────────────
ResEq y1  y2  y3
```

```
exp=y3
 exp=0
•H=18.000000000001
 bound={-1ε99,1ε99}
•left-rt=2ε-11

GRAPH WIND ZOOM TRACE SOLVE
```

NOTE: You need to closely examine the equation in y3 and see what type of function it represents. In this case, y3 contains H to no power higher than one, so it is a linear equation and has only one root. If y3 contains the variable squared, the equation is quadratic and you need to try different guesses because there could be two solutions, and so forth.

Return to the home screen and press ALPHA ([(H) ENTER to check that the TI-86 knows that the answer from the SOLVER is a value of H. Next, find *w* by evaluating y2. The TI-86 does not know that the output of y2 equals W, so be sure to store this value in W.

```
H
                    18
y2
                    18
Ans→W
                    18
```

The last thing to do is to find the output $V(h, w)$ at the current values of *h* and *w*. To do this, call up y1 on the home screen.

```
y1
                 11664
```

The critical point has coordinates $h = 18$ inches, $w = 18$ inches, and $V = 11,664$ cubic inches.

10.2.1b CLASSIFYING CRITICAL POINTS USING THE DETERMINANT TEST

Once you find one or more critical points, the next step is to classify each as a point at which a maximum, a minimum, or a saddle point occurs. The Determinant Test often will give the answer. Also, because this test uses derivatives, the calculator's numerical derivative der1 can help.

We illustrate with the critical point that was found in Section 10.2.1a of this *Guide*. To use the Determinant Test, we need to calculate the four second partial derivatives of *V* and then evaluate them at the critical point values of *h* and *w*.

Enter the functions V in y1, V_h in y2 and V_w in y3. These quantities are given on page B-93 of this *Guide*.

H and W should contain the <u>unrounded</u> values of the inputs at the critical point. (Here, H and W are integers, but this will not always be the case.)

```
Plot1 Plot2 Plot3
\y1⊟108H*W-2H²*W-2H*…
\y2⊟108W-4H*W-2W²
\y3⊟108H-2H²-4H*W
────────────────────
x(x)² WIND ZOOM TRACE GRAPH
  x   y   INSf DELf SELCT▶
```

```
H
                    18
W
                    18
```

Take the derivative of y2 = V_h with respect to h and we have V_{hh}, the partial derivative of V with respect to h and then h again. Find $V_{hh} = -72$ at the critical point.	```der1(y2,H,H)``` -72 ```der1(y3,H,H)``` -36
Take the derivative of y3 = V_w with respect to h and we have V_{wh}, the partial derivative of V with respect to w and then h. Find $V_{wh} = -36$ at the critical point.	If you prefer, enter the numerical value of H in the 3rd position.
Take the derivative of y2 = V_h with respect to w and we have V_{hw}, the partial derivative of V with respect to h and then w. Find $V_{hw} = -36$ at the critical point. (*Note:* V_{wh} must equal V_{hw}.)	```der1(y2,W,W)``` -36 ```der1(y3,W,W)``` -72
Take the derivative of y3 = V_w with respect to w and we have V_{ww}, the partial derivative of V with respect to w and then w again. Find $V_{ww} = -72$ at the critical point.	If you prefer, enter the numerical value of W in the 3rd position.
The second partials matrix is $\begin{bmatrix} V_{hh} & V_{hw} \\ V_{wh} & V_{ww} \end{bmatrix}$. Find the value of $D = (-72)(-72) - (-36)^2 = 3888$. Because $D > 0$ and $V_{hh} < 0$ at the critical point, the Determinant Test tells us that $(h, w, V) = (18, 18, 11664)$ is a relative maximum point.	```-72*-72-(-36)²``` 3888

NOTE: The values of the second partial derivatives were not very difficult to determine without the calculator in this example. However, with a more complicated function, we strongly suggest using the above methods to provide a check on your analytic work to avoid making simple mistakes. You can also use the function in y1 to check the derivatives in y2 and y3. We use some randomly chosen values of S and M to illustrate this quick check rather than use the values at the critical point, because the critical point values give 0 for y2 and y3.

Store different values in H and W and evaluate y2 at these values. Then evaluate the TI-86's derivative of y1 with respect to H at these same values.
 	
Evaluate y3 at these stored values of H and W. Then evaluate the TI-86's derivative of y1 with respect to W at these same values.
	```8.57→H:19.72→W```
	    19.72
	```y2```
	  676.0016
	```der1(y1,H,8.57)```
	  676.0016
	`evalF` `nDer` `der1` `der2` `fnInt`
		```8.57→H:19.72→W```
		    19.72
		```y3```
		  102.6686
		```der1(y1,W,19.72)```
		  102.6686
		`evalF` `nDer` `der1` `der2` `fnInt`

10.2.2 **FINDING CRITICAL POINTS USING MATRICES** To find the critical point(s) of a smooth, continuous multivariable function, we first find the partial derivatives with respect to each of the input variables and then set the partial derivatives equal to zero. This gives a *system of simultaneous equations* to be solved. Here we illustrate solving this linear system using an orderly array of numbers that is called a *matrix*. The TI-86 simplifies this solving process using matrices with its *simultaneous equation solver*.

WARNING: This solution method applies only to *linear* systems of equations. That is, the system of equations should not have any variable appearing to a power higher than 1 and should not contain a product of any variables. If the system is not linear, you must use algebraic solution methods to solve the system. Using your TI-86 with the algebraic solution method is illustrated in Section 10.2.1a of this *Guide*.

Consider the cake volume index function in Example 1 of Section 10.2 in *Calculus Concepts*:

$$V(l, t) = -3.1l^2 + 22.4l - 0.1t^2 + 5.3t$$

When *l* grams of leavening is used and the cake is baked at 177ºC for *t* minutes. The system of equations derived from the partial derivatives of *V* are

$$V_l = -6.2l + 22.4 = 0$$
$$V_t = -0.2t + 5.3 = 0$$

Even though these 2 linear equations can easily be solved for *l* and *t*, we use them to illustrate the matrix solution method. When using this method, you MUST remember to write the equations so that the constant terms are on the right-hand side of the equations and the coefficients of the input variables occupy the same positions on the left-hand side of each equation:

$$-6.2l + \quad 0t = -22.4 \qquad\qquad [1]$$
$$0l + -0.2t = -5.3 \qquad\qquad [2]$$

Access the simultaneous equation solver with 2nd TABLE (SIMULT).

| Enter 2 for the number of equations. You next see a screen with the letters *a*, *b*, and *x* with some subscripts. | ```SIMULT Number=2■``` | ```a1,1X1+a1,2X2=b1 a1,1=■ a1,2= b1=``` ‾PREV‾NEXT‾CLRa‾‾‾‾SOLVE‾ |

The subscripts give the row and column that the coefficients in the equations appear. The *a*s represents the coefficients on the left-hand side of the equation and *b* represents the constant term on the right-hand side of the equation. The first value in the subscript stands for the equation and the second number stands for the column in which the variable appears. For instance, "$a_{1,2}$" refers to equation 1 and the coefficient of the second variable, *t*. Note that x_1 refers to the input variable in the first position (our *l*) and x_2 refers to the input variable in the second position (our *t*).

Enter the coefficients and constants in equations 1 and 2 above, reading from left to right. Press ENTER or F2 [NEXT] to go to the screen for the second equation.	```a1,1X1+a1,2X2=b1 a1,1=-6.2 a1,2=0 b1=-22.4``` ‾PREV‾NEXT‾CLRa‾‾‾SOLVE‾	```a2,1X1+a2,2X2=b2 a2,1=0 a2,2=-.2 b2=-5.3``` ‾PREV‾NEXT‾CLRa‾‾‾SOLVE‾
If you want to return to the screen for the first equation, press F1 [PREV]. Press F5 [SOLVE] to solve the system.	```x1=3.61290322581 x2=26.5``` ‾COEFS‾STOa‾STOb‾STOx‾	Carefully notice the order in which you wrote the equations and see that $x_1 = l$ and $x_2 = t$.
We need to use these unrounded values to find the multivariable function output at the critical point. To do this, have the cursor to the equals sign of $x_1 = h$ and press STO▸ 7 [L] ENTER		```x1=3.61290322581 x2=26.5 Sto L``` ‾COEFS‾STOa‾STOb‾STOx‾

Note: If you do not store the unrounded solution values to the correct variable names, you will not obtain the correct number for the optimal value of the function at the critical point.

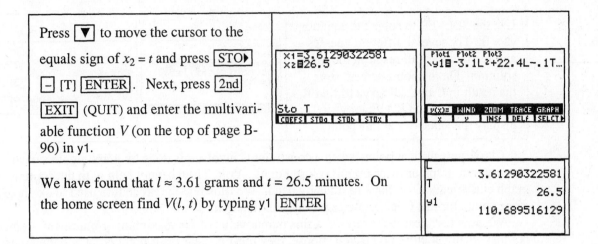

Press ▼ to move the cursor to the equals sign of $x_2 = t$ and press STO▶ [–] [T] ENTER. Next, press 2nd EXIT (QUIT) and enter the multivariable function V (on the top of page B-96) in y1.	
We have found that $l \approx 3.61$ grams and $t = 26.5$ minutes. On the home screen find $V(l, t)$ by typing y1 ENTER	

10.3 Optimization Under Constraints

Optimization techniques on your calculator when a constraint is involved are the same as the ones discussed in Sections 10.2.1a and 10.2.2 except that there is one additional equation in the system of equations to be solved.

10.3.1 FINDING OPTIMAL POINTS ALGEBRAICALLY AND CLASSIFYING CRITICAL POINTS UNDER CONSTRAINED OPTIMIZATION

We illustrate solving a constrained optimization problem with the functions given in Example 1 of Section 10.3 – the Cobb-Douglas production function $f(L, K) = 48.1L^{0.6}K^{0.4}$ subject to the constraint $g(L, K) = 8L + K = 98$ where L worker hours (in thousands) and $\$K$ thousand capital investment are for a mattress manufacturing process.

We first find the critical point(s). Because this function does not yield a linear system of partial derivative equations, we use the algebraic method. We employ a slightly different order of solution than that shown in the text. The system of partial derivative equations is

$$\left.\begin{array}{l} 28.86L^{-0.4}K^{0.4} = 8\lambda \\ 19.24L^{0.6}K^{-0.6} = \lambda \\ 8L + K = 98 \end{array}\right\} \Rightarrow \begin{array}{l} 28.86L^{-0.4}K^{0.4} = 8(19.24L^{0.6}K^{-0.6}) \qquad [4] \\ \\ K = 98 - 8L \qquad\qquad\qquad\qquad\qquad [5] \end{array}$$

Equation 4 was derived by substituting λ from the second equation on the left into the first, and equation 5 was derived by solving the third equation on the left for K. We now solve this system of 2 equations (equations 4 and 5) in 2 unknowns (L and K) using the methods shown in Section 10.2.1a.

Clear the y(x)= list. Enter the function f in y1 and the expression for K in y2. Rewrite the *other* equation (equation 4) so that it equals 0, and enter the non-zero side in y3. Substitute y2 into y3. (See the note below.)	

NOTE: Remember that K = y2. Put the cursor on the first K in y3 and replace K by y2. (You need to use the 2nd DEL (INS) key to type "2" in y2.) Do the same for the other K in y3.

The expression now in y3 is the left-hand side of an equation that equals 0 and contains only one variable, namely L. (We are not sure how many answers there are to this equation.)

| Use the SOLVER to solve the equation $y3 = 0$. Try several different guesses and see that they all result in the same solution. On the home screen, store the result in L and call up y2 to find K. (Store this value in K.) Enter y1 to display the value of f at this point. | exp=y3
  exp=0
∎L=7.3500000000001
bound={-1ᴇ99,1ᴇ99}
∎left-rt=0
[GRAPH][WIND][ZOOM][TRACE][SOLVE] | y2   7.35
Ans→K   39.2
           39.2
y1
   690.608479798 |

Classifying critical points when a constraint is involved is done by graphing the constraint on a contour graph or by examining close points. Your TI-86 cannot help with the contour graph classification – it must be done by hand. We illustrate the procedure used to examine close points for this Cobb-Douglas production function.

We now test close points to see if this output value of f is maximum or minimum. Remember that whatever close points you choose, they must be near the critical point and they must be on the constraint g.

WARNING: Do not round during this procedure. Rounding of intermediate calculations and/or inputs can give a false result when the close point is very near to the optimal point.

| Choose a value of L that is less than $L = 7.35$, say 7.3. Find the value of K so that $8L + K = 98$. Remember that K = y2, so store 7.3 in L and call up y2. Store this value in K. Then find the value of f at L = 7.3, K = 39.6.

At this close point, the value of f is less than the value of f at the optimal point (690.6084798...). | 7.3→L:y2   39.6
Ans→K
          39.6
y1
   690.584564125 |
| Choose another value of L, this time one that is more than $L = 7.35$, say 7.4. Find the value of K so that $8L + K = 98$ by storing 7.4 in L and calling up y2. Then find the value of f at L = 7.4, K = 38.8.

At this close point, the value of f is less than the value of f at the critical point (690.6084798...). Thus, (7.35, 39.2, 690) is a maximum point for mattress production. | 690.584564125
7.4→L:y2
         38.8
Ans→K
         38.8
y1
   690.584455414 |

NOTE: Because the constraint is the equation in y2, when we call up y2 in the procedure above we are finding points *on* the constraint g. If you have the constraint in a different location, you need to use the constraint location in the y(x)= list to find the value of K.

10.3.2 FINDING OPTIMAL POINTS USING MATRICES AND CLASSIFYING CRITICAL POINTS UNDER CONSTRAINED OPTIMIZATION

Remember that a matrix solution with the TI-86's simultaneous equation solver only can be used with a system of linear partial derivative equations. We choose to illustrate this matrix process of solving (discussed in Section 10.2.2) for the sausage production function in Example 2 of Section 10.3 in *Calculus Concepts*. The system of partial derivative equations and the constraint, written with the constant terms on the right-hand side of the equations and the coefficients of the three input variables in the same positions on the left-hand side of each equation, is

$$-5.83s - \lambda = -1.13$$
$$-5.83w \qquad -\lambda = -1.04$$
$$w + s = 1$$

(The order in which the variables appear does not matter as long as it is the same in each equation.)

When a particular variable does not appear in an equation, its coefficient is zero. Access the simultaneous linear equation solver with [2nd] [TABLE] (SIMULT).

Enter 3 as the number of equations.
 	
Enter the coefficients of $x_1 = w$, $x_2 = s$, and $x_3 = \lambda$ in the a positions and the constant terms in the b positions for the first two equations.
	`a1,1x1...a1,3x3=b1`
	`a1,1=0`
	`a1,2=-5.83`
	`a1,3=-1`
	`b1=-1.13`
	
	`PREV│NEXT│CLRq│    │SOLVE│`
		`a2,1x1...a2,3x3=b2`
		`a2,1=-5.83`
		`a2,2=0`
		`a2,3=-1`
		`b2=-1.04`
		
		`PREV│NEXT│CLRq│    │SOLVE│`

Repeat the process the third equation.
 	
Then press F5 [SOLVE].
	`a3,1x1...a3,3x3=b3`
	`a3,1=1`
	`a3,2=1`
	`a3,3=0`
	`b3=1`
	
	`PREV│NEXT│CLRq│    │SOLVE│`
		`x1=.492281303602`
		`x2=.507718696398`
		`x3=-1.83`
		
		`COEFS│STOq│STOb│STOx│`

Store these results in W and S for use with what follows.	`x1=.492281303602`
	`x2=.507718696398`
	`x3=-1.83`
	
	`Sto W`
	`COEFS│STOq│STOb│STOx│`
		`x1=.492281303602`
		`x2=.507718696398`
		`x3=-1.83`
		
		`Sto S`
		`COEFS│STOq│STOb│STOx│`

(See a more detailed discussion of storing the results on pages B-96, B-97 of this *Guide*.)

Because a contour graph is given (See Figure 10.24 in the text), it is easiest to use it to verify that the optimal point is a minimum. However, we show the method of choosing close points on the constraint to give another example of this method of classification of optimal points.

First, enter the function P with inputs w and s into y1. Place the constraint, solved for s, in y2. Next, substitute the unrounded values of w and s into the function to find the optimal value.	`Plot1 Plot2 Plot3`
	`\y1█10.65+1.13W+1.04...`
	`\y2█1-W`
	
	`y(x)=│WIND│ZOOM│TRACE│GRAPH`
	`  x  │  y  │INSf│DELf│SELCT▶`

	`W`
	`         .492281303602`
	`S`
	`         .507718696398`
	`y1`
	`         10.2771526587`

Choose a value of W that is less than 0.492, say 0.48. Store 0.48 in W and call up y2 to find the value of S. Store this value in S. Then find the value of f at W = 0.48, K = 0.52.
 	
At this close point, the value of P is more than the value of P at the critical point.
	`.48→W`
	`                      .48`
	`y2→S`
	`                      .52`
	`y1`
	`              10.278032`

Now choose a value of W that is more than 0.492, say 0.51. Store 0.51 in W and call up y2 to find the value of S. Store this value in S. Then find the value of f at W = 0.48, K = 0.49.
 	
At this close point, the value of P is more than the value of P at the critical point.
	`.51→W`
	`                      .51`
	`y2→S`
	`                      .49`
	`y1`
	`              10.278983`

Because the value of P at the close points is more than its value at the critical point, $w \approx 0.492$ whey protein, $s \approx 0.508$ skim milk powder, and $P \approx 10.277\%$ cooking loss is a minimum point.

Chapter 11　Dynamics of Change:　Differential Equations and Proportionality

The TI-86 offers many opportunities when studying differential equations. It draws slope field graphs and has as built-in commands many of the techniques that must be programmed into other calculators. However, in this *Guide*, we look only at those tools that are necessary for the study of differential equations as presented in Chapter 11 of *Calculus Concepts*. We encourage you to further explore the TI-86 capabilities by reading and working through pages 132-150 of your *TI-86 Graphing Calculator Guidebook*.

 ## 11.3　Numerically Estimating by Using Differential Equations:　Euler's Method

Many of the differential equations we encounter have solutions that can be found by determining an antiderivative of a given rate-of-change function. Thus, many of the techniques that we learned using the TI-86's numerical integration function apply to this chapter. (See Chapter 6 of this *Guide*.)

11.3.1　EULER'S METHOD FOR A DIFFERENTIAL EQUATION WITH ONE INPUT
VARIABLE　You may encounter a differential equation that cannot be solved by standard methods and you may need to draw an accumulation graph for a differential equation without first finding an antiderivative. In either of these cases, numerically estimating a solution using Euler's method is helpful. This method relies on the use of the derivative of a function to approximate the change in that function. Recall from Section 5.1 of *Calculus Concepts* that the approximate change in a function f at a point is the rate of change of f at that point times a small change in x. That is,

$$f(x + h) - f(x) \approx f'(x) \cdot h \text{ where } h \text{ represents the small change in } x.$$

Now, if we let $b = x + h$ and $x = a$, the above expression becomes

$$f(b) - f(a) \approx f'(a) \cdot (b - a) \quad \text{or} \quad f(b) \approx f(a) + (b - a) \cdot f'(a)$$

The starting values for the coordinates of the point (a, b) will be given to you and are often called the *initial condition*. The next step is to repeatedly apply the formula given above to use the slope of the tangent line at $x = a$ to approximate the change in the function between the inputs a and b. When h, the distance between a and b, is fairly small, Euler's method will often give close numerical estimates of points on the solution to the differential equation containing $f'(x)$.

　　We illustrate Euler's method for a differential equation containing one input variable with the differential equation in Example 1 of Section 11.3. This equation gives the rate of change of the total sales of a computer product t years after the product was introduced:

$$\frac{dS}{dt} = \frac{6.544}{\ln(t + 1.2)} \text{ billion dollars per year}$$

Because Euler's method involves a repetitive process, a program that performs the calculations used to find the approximate change in the function can save you time and eliminate computational errors and some error in rounding.

Before using this program, you must have the differential equation in the y1 location of the y(x)= list with x as the input variable. From the home screen, access the Euler program with the keys `PRGM` `F1` [NAMES].	

WARNING: Be wary of the fact that there is some error involved in each step of the Euler approximation process that is compounded when each result is used to obtain the next result.

The code for program EULR is listed in the *TI-86 Program Appendix*. Run the program. Each time the program stops for input or for you to view a result, press ENTER to continue.

We choose to use 16 steps. Enter this value. The interval is 4 years, so enter the step size = $\dfrac{\text{length of interval}}{\text{number of steps}} = \dfrac{4}{16} = 0.25$.	`HAVE dy/dx in y1`
	`Number of steps= 16`
	`Step size= 0.25`■	

The initial condition is given as the point (1, 53.2). Enter these values when prompted for them.	`Number of steps= 16`
	`Step size= 0.25`
	`Initial input= 1`
	`Initial output= 53.2`
	`Input,Output is`
	`              1.25`
	`    55.2749378245`
		The first application of the formula gives an estimate for the value of the total sales at $x = 1.25$:
		
		$S(1.25) \approx 55.275$

Press ENTER several more times to obtain more estimates for total sales.
 	
Record the input values and output estimates on paper as the program displays them.
	`      57.1006513204`
	`Input,Output is`
	`              1.75`
	`     58.747766427`
	`Input,Output is`
	`                 2`
	`    60.2600535235`
		Continue pressing the ENTER key to obtain more estimated values of points on the total sales function S.

When 16 steps have been completed (that is, after the input reaches 5), the program draws a graph of the points *(input, output estimate)* connected with straight line segments. This graph is an estimate of the graph of the differential equation solution.	`      71.7020961191`
	`Input,Output is`
	`              4.75`
	`    72.6420741685`
	`Input,Output is`
	`                 5`
	`    73.5594275692`
		This is an estimate of the graph of the function $S(t)$.

In the y(x)= list, turn Plot3 off, turn y1 on, and press EXIT F3 [ZOOM] MORE F1 [ZFIT] to draw the graph of the differential equation.	`Plot1 Plot2 Plot3`
	`\y1■6.544/ln (x+1.2)`
	
	`y(x)= WIND ZOOM TRACE GRAPH`
	`  x    y   INSf DELf SELCT▶`
		This is the slope graph – the graph of $S'(t)$.

11.3.2 EULER'S METHOD FOR A DIFFERENTIAL EQUATION WITH TWO INPUT VARIABLES
Program EULR can be used when the differential equation is a function of x and y with $y = f(x)$. Follow the same process that is illustrated in the previous section of this *Guide*, but enter $\dfrac{dy}{dx}$ in y1 using the letters x and y as they are written in the given equation.

If the differential equation is written in terms of variables other than x and y, let the derivative symbol be your guide as to which variable corresponds to the input and which corresponds to the output. For instance, if the rate of change of a quantity is given by $\dfrac{dP}{dn} = 1.346P(1 - n^2)$, compare $\dfrac{dy}{dx}$ to $\dfrac{dP}{dn}$, entering in y1 the expression $1.346y(1 - x^2)$. The differential equation may be given in terms of y only. For instance, if $\dfrac{dy}{dx} = k(30 - y)$ where k is a constant, enter y1 = K(30 − y). Of course, you need to store a value for k or substitute a value for k in the differential equation before using program EULR. It is always better to store the exact value for a constant instead of using a rounded value.

WARNING: When using program EULR, you must use a lower-case *y* when entering an equation involving two variables in the y(x)= list. Use 2nd ALPHA (alpha) 0 (y) to type *y*.

We illustrate using Euler's method with two input variables by using the equation in Example 2 of Section 11.3 in *Calculus Concepts*.

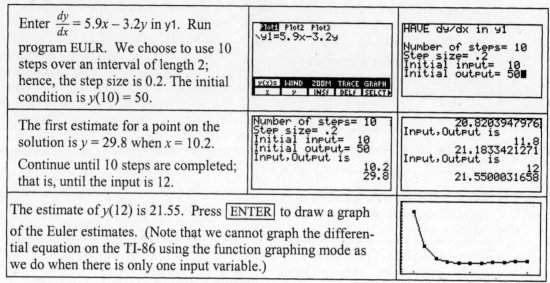

Enter $\frac{dy}{dx} = 5.9x - 3.2y$ in y1. Run program EULR. We choose to use 10 steps over an interval of length 2; hence, the step size is 0.2. The initial condition is $y(10) = 50$.	
The first estimate for a point on the solution is $y = 29.8$ when $x = 10.2$. Continue until 10 steps are completed; that is, until the input is 12.	
The estimate of $y(12)$ is 21.55. Press ENTER to draw a graph of the Euler estimates. (Note that we cannot graph the differential equation on the TI-86 using the function graphing mode as we do when there is only one input variable.)	

Troubleshooting the TI-86

❑ CALCULATOR BATTERIES

- *Will I lose all my programs when I change batteries?*

There is no guarantee that changing batteries will not reset your calculator's memory. However, a method that seems to work well is to make sure your calculator is turned off, take the old batteries out one at a time, replacing each with a new battery before removing the next. Be sure you have the direction of the batteries correct (+ and – alternating).

- *What is the backup battery?*

The backup battery is a round battery above the AAA batteries and under a piece of plastic that is fastened to the calculator with a small screw.

- *How long will my batteries last?*

How long your batteries last depends on how much you use your calculator. The four AAA batteries in the calculator usually last for about a year with normal use in a calculus course. The backup battery can last anywhere from 2 to 5 years.

- *When do I need to replace the batteries?*

You need to replace the AAA batteries if your screen is not dark enough and 9 is the value shown in the upper right hand column when you press **2nd**, release that key, and then press and hold the up arrow. (See the second bulleted item under the GENERAL category.)

- *I've replaced the four AAA batteries and my calculator still won't come on. What next?*

Sometimes when you replace batteries, the calculator resets to its lightest screen setting. Darken the display according to the directions given in the second bulleted item under the GENERAL category. If you have replaced the AAA batteries, you have darkened the display, and your calculator still will not turn on, it is likely that the backup battery needs replacing.

❑ DELETING PROGRAMS OR OTHER INFORMATION FROM THE TI-86

- *How do I delete information from my calculator?*

Press **2nd 3 [MEM] F2 [Delete]** and then press the F-key corresponding to the type of item you are deleting. For instance, to delete a program, press **MORE F5 [Prgm]**. Use the down arrow to move the cursor next to the information you want to delete and press **ENTER**. **EXIT** returns you to the home screen.

- *When I try to run a program, I keep getting ERROR 10 DATA TYPE. What can I do?*

You probably stored a picture or graph data base to a variable name that conflicts with a variable name used by the program. Press **2nd 3 [MEM] F2 [Delete] F1 [ALL]**, scroll down the list of variables and see if you have one that says GDB (graph data base) or PIC (picture). If you find one, and it is not something you want to keep, have the cursor next to the name and delete it by pressing **ENTER**. (Be very careful on this screen – pressing **ENTER** deletes whatever the cursor is next to and there is no undo instruction!)

If the **PIC** or **GDB** is something you do want to save, return to the home screen with **EXIT**, press **GRAPH MORE RECDB** (recall graph data base) or **GRAPH MORE MORE RCPIC** (recall picture) and type in the current name of your picture or graph data base, press **ENTER** and then either **STPIC** or **STGDB** (you may have to press **MORE** to find these). Type in the new name and press **ENTER**. Be sure that the new name has more than a single letter! Return to the delete screen (see the first paragraph) and delete the old graph data base or picture. This should correct the problem.

B-103

❑ GENERAL INFORMATION

- *Why do my calculator screens not look like the ones in this Guide?*

Make sure your calculator settings are as specified in the Setup Instructions on page *B*-1.

- *How do I make the display screen darker or lighter?*

To make the display darker, press **2nd**, release that key, and then press and hold the up arrow until the display is dark enough. To make the display lighter, press **2nd**, release that key, and then press and hold the down arrow until the display is light enough. You should see a number in the upper right corner of the display screen that varies between 0 (light) to 9 (dark.)

- *I get tired of pressing so many keys to have to find instructions that I use all the time. Is there something I can do to make the process simpler?*

Yes, you can access the instructions in the **CATALOG** (See page *B*-36 of this *Guide*) or the TI-86 has a **CUSTOM** menu in which you can enter commands that you use many times. See pages 44-45 of the *TI-86 Graphing Calculator Guidebook*.

- *Is there a difference in lower-and upper-case letters on the TI-86?*

Yes, the TI-86 makes a distinction in the case. For instance, *P* and *p* are different variable names on this calculator. Also, the TI-86 considers *PT* as the name of a single variable, but *P*T* or *P*(*T*) means the multiplication of *P* and *T*.

- *What causes an ERROR 14 UNDEFINED message?*

Anytime you use a variable consisting of one or more letters, it must first be defined by storing a value to that variable. For example, to use *P* in a formula, you must first store some value to *P*. Another time that this error occurs (especially in Chapters 9 and 10 of *Calculus Concepts*) is when you forget to put a times sign in between two letters that each stand for different variables. For instance, if you enter in the graphing list (or the solver) the equation y1 = 2A + B^2 + 3AB, the TI-86 thinks that there are three inputs: A, B, and AB. This equation should be entered on one line as y1 = 2A + B^2 + 3A*B.

- *I have either lost or never got the owner's manual for the TI-86. What can I do?*

The *TI-86 Graphing Calculator Guidebook* is available at the Texas Instruments web site with address address **www.TI.com/calculators** .

- *I can't find the information I need in this Guide.*

See the material entitled **How to Use This Guide** following the *Contents* on page vi.

❑ GRAPHING

- *Why don't I see a graph when I press the correct keys to make it draw?*

Make sure that you have the function entered in the y(x)= (graphing) list, using a *lower-case x* for the input variable. (See Sections 1.1.1a, b.) Also check that your function is turned on. (See Section 1.2.2b.) If the function is turned on but you still cannot see the graph, check your window settings – maybe the function did graph, but it graphed outside your window.

- *I've lost y1 (and/or y2, etc.) in the graphing list. How do I get them back?*

If you don't need any of the current functions in the list, press **DELf** on the graphing menu until it resets to y1. If you want to save the functions that are there, place the cursor on the line above which you want to insert a function location. Press **INSf** on the graphing menu.

- *What do I do if a get a strange-looking graph or no graph instead of a scatter plot of data when I press* **GRAPH F3 [ZOOM] MORE F5 [ZDATA]**?

The scatter plot setup has somehow been changed, is not correctly set, or is not turned on. Refer to Section 1.4.2a of this *Guide* for instructions.

- *While trying to draw the graph of an equation that I entered in the* y(x)= *list, I get an* ERROR 13 DIMENSION. *What do I do?*

The TI-86 will not draw the graph of a function when one of the scatter plots is turned on and the lists are of unequal length or contain no data. To correct this problem, choose **1: [Quit]** and turn off all stat plots by making sure that the names (Plot 1, Plot 2, Plot 3) are not dark. The graph should now draw with out any problems.

- *Do I have to clear the function location in the* y(x)= *list before pasting in another function when finding an equation of best-fit?*

Most of the time, it is not necessary to first clear any previously-entered function from the chosen location of the **y(x)=** list. However, if you receive an error message when finding the equation, clear the desired function location and press **2nd ENTER (ENTRY) ENTER** to recall and repeat the regression instruction. If you still obtain an error message, reset the statistical setup as described on the first page of this *Guide*.

❑ LISTS

- *I don't have all the lists when I press* **2nd + (STAT) F2 [EDIT]**. *How do I fix this?*

Make sure your calculator settings are as specified in Figure 2 using the second bulleted item in the list describing the Setup Instructions on the first page of this *Guide*.

❑ PROGRAMS

- *Where can I get the programs that are used in this Guide?*

The programs can be transferred to your calculator from another student or your teacher's calculator and a link chord. If no one has the programs, they can be downloaded from the *Calculus Concepts* CD-ROM or web site. A TI Graph-Link™ cable and the linking software for your particular calculator are needed to transfer the programs from the Internet to a computer. A special cable is needed to transfer the programs from the computer to your calculator. The linking software is free at the TI web site with address **www.ti.com/calculators**. If you cannot find a cable, ask at the learning center for your school, check with your instructor, or search the TI web site for instructions on how to purchase one.

- *Why won't a program run, or why do I get an error when I try to run a program?*

You may have unknowingly deleted or altered one or more lines of the program code. The easiest method of fixing the problem is to use the link chord to re-transfer the entire program into your calculator. If this is not possible, you can find the correct program code in the *TI-86 Program Appendix* at the *Calculus Concepts* web site. Press **PRGM right arrow [EDIT]** and **down arrow** until the name of the corrupt program is highlighted and press **ENTER**. Compare the code in your calculator with what is printed in the *TI-86 Program Appendix* until you find the error and correct it by retyping the proper code. Consult the *TI-86 Graphing Calculator Guidebook* that came with your calculator for the location of the symbols in the program code.

- *Program NUMINTGL keeps giving an error when I try to run it. What can I do?*

Delete all functions from the y(x)= list (except the function in y1) before using the program named NUMINTGL. This program uses every memory location in the calculator and all the variables have to be defined before the program will run. Additionally, single-letter names cannot be used to name other types of stored information. (See the fifth bulleted item under GENERAL INFORMATION. See also DELETING PROGRAMS OR OTHER INFORMATION FROM THE TI-86 on page B-103 of this *Guide*.)

TI-86 Program Appendix

The programs listed below are referenced in Part *B* of the *Graphing Calculator Instruction Guide* for *Calculus Concepts*. They should be transferred to your TI-86 via a cable by using the LINK mode and another TI-86 calculator or by using the TI-GRAPH LINK™ cable and GRAPH LINK™ software for a PC or Macintosh computer that is accessible through the site **www.education.ti.com**. These programs are available for download on the *Calculus Concepts* CD-ROM and web site. As a last resort, the programs can be typed into your calculator. Please refer to your *TI-86 Graphing Calculator Guide-book* for instructions on entering the programs or transferring them via cable from another TI-85 or TI-86. The program code follows.

Program Name	Program Size (bytes)	Chapter first referenced
DIFF	891	1
NUMINTGL	1536	6
EULR	334	11
LSLINE*	526	1
SECTAN*	583	3

*These programs are more for instructional exploration than for use in working problems.

```
DIFF          • Program
:ClLCD
:Disp "Store x-values in L1"
:Disp "Store y-values in L2"
:Disp "Continue?"
:Menu(1,"Yes",GO,2,"Quit",NO)
:Lbl NO
:Stop
:Lbl GO
:dimL L1→M
:dimL L2→N
:If M≠N:Goto Z
:N-1→dimL L6
:For(H,1,N-1,1)
:L1(H+1)-L1(H)→L6(H)
:End
:For(H,1,N-2,1)
:If L6(H+1)≠L6(H)
:Goto FN
:End
:N-1→dimL L3
:For(A,1,N-1,1)
:L2(A+1)-L2(A)→L3(A)
:End
:N-2→dimL L4
:For(B,1,N-2,1)
:L3(B+1)-L3(B)→L4(B)
:End:ClLCD
:Disp "Choice?"
:Lbl RC
:Menu(1,"1st",FST,2,"2nd",SEC,3,"%",PE
R,4,"Quit",QT)
:Lbl FST
:Disp "1st differences in L3"
:Disp L3
```

```
(Program DIFF continued)
:Goto RC
:Lbl SEC
:Disp "2nd differences in L4"
:Disp L4
:Goto RC
:Lbl PER
:N-1→dimL L5
:1→E
:For(E,1,N,1)
:If L2(E)==0:Goto W
:End
:For(E,1,N-1,1)
:(L3(E)/L2(E))*100→L5(E)
:End
:Disp "percent diffs in L5"
:Disp L5
:Goto RC
:Lbl Z
:Disp "Lists are of unequal"
:Disp "length. Check data."
:Stop
:Lbl W
:0→dimL L5
:Disp "percent diffs not"
:Disp "calculated...cannot"
:Disp "divide by 0"
:Goto RC
:Lbl QT
:Stop
:Lbl FN
:Disp "Input values not"
:Disp "evenly spaced"
:Stop
```

```
NUMINTGL        • Program
:ClLCD
:Disp "Enter f(x) in y1"
:Disp ""
:Disp "Continue?"
:Menu(1,"Yes",YS,2,"No",NO)
:Lbl NO:Stop
:Lbl YS
:Disp ""
:Disp "Draw Pictures?"
:Menu(1,"Yes",YE,2,"No",NR)
:Lbl YE:1→H:Goto LE
:Lbl NR:2→H
:Lbl LE
:ClLCD
:Input "Left endpoint? ",A
:Input "Right endpoint? ",B
:If H==1:Then
:A→xMin:B→xMax
:iPart ((B-A)/20)→W
:If W==0:0.1→W
:seq(V,V,A,B,W)→L5
:dimL L5→N:N→dimL L6
:For(j,1,N,1)
:L5(j)→x:y1→L6(j)
:End
:real L6→L6
:min(L6)→yMin
:If yMin>0:0→yMin
:max(L6)→yMax
:If yMax<0:0→yMax
:W→xScl
:iPart ((yMax-yMin)/10)→yScl
:0→dimL L5:0→dimL L6
:End
:Lbl A0
:ClLCD
:Disp "Enter Choice:"
:Disp "Left Rect  (1)"
:Disp "Right Rect (2)"
:Disp "Trapezoids (3)"
:Input "Midpt Rect (4) ",R
:Lbl A1
:ClDrw
:Disp ""
:Input "N? ",N
:(B-A)/N→W
:0→S:1→C
:Lbl A2
:If R==1:Goto A3
:If R==2:Goto A4
:If R==3:Goto A3
:If R==4:Goto A5
:Lbl A3
:A+(C-1)W→x
:x→J:x+W→L
:Goto A7
:Lbl A4
:A+C*W→x
:x-W→J:x→L
:Goto A7
:Lbl A5
```

(Program NUMINTGL continued)

```
:If H≠1:Then
:If N>5:Then
:1→Z:W/2→H:A→x
:Lbl A8
:x+H→x:y1+S→S
:A+Z*W→x
:IS>(Z,N):Goto A8
:S*W→S:Goto T
:End:End
:A+C*W-W/2→x
:x-W/2→J
:x+W/2→L
:Goto A7
:A→G:G+W→G:G→V
:Lbl A9
:V→x:y1→Y:V+W→x
:4Y+2y1+S→S
:V+2*W→V
:If V<B:Goto A9
:G-W→x:y1→E
:B→x:y1→F
:(W/3)*(S+E-F)→S
:Goto T
:Lbl A7
:y1→K:K+S→S
:If H==1:Goto D
:Lbl I
:IS>(C,N)
:Goto A2
:If R==3:Then
:A→x:y1→P
:B→x:y1→Q
:S+(Q-P)/2→S
:End
:W*S→S
:Lbl T
:Disp "SUM=",S
:Pause
:ClLCD
:Lbl E
:ClLCD
:Disp "Enter Choice:"
:Disp "Change N       (1)"
:Disp "Change Method (2)"
:Input "Quit          (3) ",T
:If T==1:Goto A1
:If T==2:Goto A0
:If T==3:Goto F
:Lbl F
:Stop
:Lbl D
:If R==3:Then
:x→I:L→x
:y1→M:I→x
:Else:K→M
:End
:Line(J,0,J,K)
:Line(J,K,L,M)
:Line(L,M,L,0)
:If C==N:Pause
:Goto I
```

LSLINE • Program
```
:FnOff
:0→A:0→B:1→C
:y1=A+B x
:L1→xStat
:L2→yStat
:dimL xStat→N
:ClLCD
:Disp "You will next view"
:Disp "the data. Use tick"
:Disp "marks to guess the"
:Disp "slope and y-intercept"
:Disp "of best fit line."
:Disp "xScl=":Outpt(6,7,xScl)
:Disp "yScl=":Outpt(7,7,yScl)
:Pause
:Lbl A1
:Scatter
:Pause
:Disp ""
:Input "slope=",B
:Input "y intercept=",A
:1→K:0→x:0→S
:Lbl V
:xStat(K)→x
:y1→Y
:(yStat(K)-Y)²+S→S
:Line(xStat(K),yStat(K),x,Y)
:K+1→K
:If K≤N
:Goto V
:Pause
:Disp "SSE=",S
:Pause
:If C==2
:Goto W
:Input "Try again? Y(1) N(2) ",C
:If C==1
:Goto A1
:LinR
:ShwSt
:Pause
:DrawF RegEq
:Pause
:a→A:b→B
:1→K:0→x:0→S
:Goto V
:Lbl W
:FnOff
```

SECTAN • Program
```
:ClLCD
:ClDrw:2→R
:Disp ""
:Disp "Have f(x) in y1 and"
:Disp "draw graph of f"
:Disp ""
:Disp "Continue? "
:Menu(1,"Yes",YS,2,"NO",NO)
:Lbl YS
:Disp ""
:Disp "x-value of point"
:Input "of tangency? ",A
```

(*Program* SECTAN *continued*)
```
:Lbl RT
:Disp ""
:Disp "Press ENTER to "
:Disp "see secant lines"
:If R==1:Goto RS
:Disp "from the left"
:Goto RU
:Lbl RS
:ClDrw
:Disp "from the right"
:Lbl RU
:Disp "approach tangent"
:Disp "line"
:Pause
:(xMax-xMin)/3→K
:If K>50:48→K
:For(J,1,5,1)
:A-K→x
:If R==1:A+K→x:x→D
:y1→B:A→x:y1→C
:(B-C)/(D-A)→M
:A→x:y1→E
:DrawF (M(x-A)+E)
:K/2→K
:End
:Pause
:If R==1:Goto RV
:1→R:Goto RT
:Lbl RV
:ClLCD
:Disp "Press ENTER to"
:Disp "see tangent line"
:Pause
:ClDrw
:TanLn(y1,A)
:Lbl NO:Stop
```

EULER • Program
```
:ClLCD
:0→dimL L1:0→dimL L2
:FnOff
:Disp "HAVE dy/dx in y1"
:Disp ""
:Input "Number of steps= ",N
:Input "Step size= ",H
:Input "Initial input=   ",x
:Input "Initial output= ",y
:For(I,1,N,1)
:x→L1(I)
:y→L2(I)
:y1→T
:x+H→x
:y+H*T→y
:Disp "Input,Output is"
:Disp x
:Disp y
:Pause
:End
:x→L1(N+1):y→L2(N+1)
:PlOff :PlOn 3
:Plot3(2,L1,L2,1)
:ZData
```